THE
DROUGHT
RESILIENT
FARM

The mission of Storey Publishing is to serve our customers by publishing practical information that encourages personal independence in harmony with the environment.

Edited by Deborah Burns
Art direction and book design by Jeff Stiefel
Text production by Erin Dawson
Indexed by Samantha Miller

Cover and interior photography by © Scott Stebner
Additional photographs by © A ROOM WITH VIEWS/
Alamy Stock Photo, 10; © Alexander62/iStock.com,
99; © andreusK/iStock.com, 66–67; © bboserup/
iStock.com, 3 top; © Blaine Harrington III/Alamy
Stock Photo, 154–155; © Brad Davison/Novel
Ways Limited, 148; © Clynt Garnham Industry/
Alamy Stock Photo, 69; courtesy of the author,
5, 23, 28, 70, 103, 126; courtesy of Amy Rickard,
158; courtesy of Green Cover Seed, 72; courtesy
of the Kansas Geological Survey, University of
Kansas, 168; courtesy of Steve Groff, Cover Crop
Consulting, 50; © DEA/G. DAGLI ORTI/Getty
Images, 29 bottom; © Design Pics Inc/Alamy
Stock Photo, 167; © Edwin Remsberg/Alamy
Stock Photo, 60; © Eugene Sergeev/Alamy Stock
Photo, 83; © Frederick M. Brown/Getty Images,
3 bottom; © Nick Moore/Alamy Stock Photo, 81;
© Perytskyy/iStock.com, 26; © Sabrina Bekeschus
Monteiro/EyeEm/Getty Images, 53; © Steffen
Hauser/botanikfoto/Alamy Stock Photo, 73;
© the_guitar_mann/iStock.com, 125; © Wayne Teel
and Lance Kearns, 63; United States Department of
Agriculture/Wikimedia Commons, viii
Illustrations by © Steve Sanford

The information in this book is true and complete
to the best of our knowledge. All recommendations
are made without guarantee on the part of the author
or Storey Publishing. The author and publisher dis-
claim any liability in connection with the use of this
information.

Storey books are available for special premium
and promotional uses and for customized editions.
For further information, please call 800-793-9396.

Storey Publishing
210 MASS MoCA Way
North Adams, MA 01247
storey.com

Printed in China by Toppan Leefung Printing Ltd.
10 9 8 7 6 5 4 3 2 1

Library of Congress Cataloging-in-Publication Data

Names: Strickler, Dale, author.
Title: The drought-resilient farm / by Dale Strickler.
Description: North Adams, MA : Storey Publishing,
    2018. | Includes bibliographical references and
    index. Identifiers: LCCN 2017061057 (print)
    | LCCN 2018000720 (ebook)
    | ISBN 9781635860030 (ebook)
    | ISBN 9781635860023 (pbk. : alk. paper)
Subjects: LCSH: Water conservation. | Droughts. | Soil
    moisture. | Livestock—Water requirements.
Classification: LCC TD388 (ebook) | LCC TD388 .S77
2018 (print) | DDC 631.4/32—dc23
LC record available at https://lccn.loc.gov/
    2017061057

THE

DROUGHT RESILIENT FARM

DALE STRICKLER

Storey Publishing

This book is dedicated to my father, Eldon, who gave me a work ethic and inspired my love of the land; my mother, Gail, who taught me to care about people other than myself; my late father-in-law, Dean Sothers, who departed this world far too early and whose help and encouragement were invaluable to me; and my family, Danell, Cole, and Cassie.

CONTENTS

PART III: LOOKING TO THE FUTURE

1

INTRODUCTION
The Ugly History of Drought, and the Potential for Drought Resilience

No natural disaster has caused as much human misery as drought. In fact, it has caused more suffering and death than all other natural disasters combined. Drought compromises our ability to acquire two of our most basic needs: water and food. Millions have suffered slow, painful deaths from thirst and starvation throughout the ages, and entire civilizations have collapsed when their supply of water or food has literally dried up.

Here are some of the great civilizations documented to have been devastated by drought.

4,200 years ago, the Akkadian Empire of Syria and the Old Kingdom of Egypt were destroyed by drought.

1,000 years later, several civilizations around the Mediterranean were devastated, including the Hittites and the Myceneans.

In the 8th century, the Mayan Empire in Central America and the Tang Dynasty in China both suffered and collapsed from drought.

In the 12th century, the Mayans again were struck, as were the Anasazi of the American Southwest, who disappeared altogether.

In the 15th century, the Ming Dynasty of China was severely affected.

In the 1930s, the most famous drought in the history of the US took place: the Dust Bowl.

All these droughts had two factors in common. The first, obviously, was lack of rainfall. The second was the destruction of natural vegetation and the substitution of a system of agriculture ill-suited for the environment.

It is often said that those who fail to heed the lessons of history are doomed to repeat the same mistakes. The Dust Bowl caused tremendous changes in our nation's environment, agricultural policy, and social fabric. Unfortunately, it was not the last drought our country has faced. A severe drought occurred in the 1950s, followed by others in 1980, 1983, 1988, 2001, 2002, 2006, 2011, 2012, and 2016.

The drought of 1980 cost the United States an estimated $56 billion, the 1988 drought cost $78 billion, and the 2012 drought was the most expensive in history. With estimated losses as high as $150 billion,

Anasazi

> Clearly the Anasazi were a very advanced civilization, as evidenced by their impressive cliff dwellings, but they disappeared altogether after a prolonged drought eliminated their ability to feed themselves.

it was the costliest environmental disaster in US history by a wide margin (in comparison, Hurricane Katrina cost an estimated $108 billion).

You would think that, given the magnitude of suffering caused by drought, finding means to reduce its impact would be a national and international priority. On the contrary, we seem simply to assume there is nothing we can do, and we choose to suffer the full effects of these disasters instead of trying to minimize them. Is there any other natural disaster that we could manage for, but do not?

Seeing Is Believing

I believe the Ken Burns documentary *The Dust Bowl* should be required viewing in every grade school. It not only vividly describes the tragedy of the Dust Bowl (using the words of the survivors to drive home the impact of the drought), but it also traces the causes of the event.

MANAGEMENT AND MISMANAGEMENT

I heard a joke in church.

A very religious man was trapped on a rooftop during a flood. A man in a rowboat came along and offered to rescue him. The man on the rooftop said, "No, thank you, but I have prayed to God and He will rescue me." So the boatman rowed away.

The flood continued to rise. A while later, a large patrol boat came up and offered to rescue the man on the rooftop. Again, he declined, saying that he had prayed to God, he was a man of faith, and that God would reward his faith by rescuing him.

The flood continued to rise. A helicopter showed up and let down a ladder. The man on the rooftop again declined, again stating that God would rescue him.

The flood continued to rise, and the man on the rooftop became submerged in water and drowned. As he entered the kingdom of heaven and came face to face with God, he asked, "I prayed and prayed to you. I was faithful to the end. Why did you not rescue me?"

God replied, "What do you mean? I sent two boats and a helicopter. I can't help it if you refused to climb aboard."

Throughout history, people have believed that drought was a punishment from a supernatural deity, and that the only possible course of action was to appease the deity through prayer or sacrifice. While it is true that there is little we can do to change the amount or timing of the rainfall we receive, there are hundreds of actions we can take that can dramatically improve how effectively precipitation is taken into the soil, is stored by the soil, and is used by vegetation. Unfortunately, most of the "management" that humans have used in agriculture throughout history has made us much more vulnerable to drought, not less. This book describes real actions we can take to make our farms and ranches much more drought-resilient. I learned these techniques through a journey of exploration that began during my childhood.

MY STORY

When I was growing up on our farm in southeast Kansas, nearly every spring it was too muddy to get our corn planted on time. Once planted, it would almost always look fantastic, due to our typically frequent rain in May and June. By July, however, the rains would diminish and the temperature would soar, and within just a few days after the last rain in July, the corn would start to wilt. Every summer, it seemed, I watched our promising crops shrivel to a fraction of their potential — and then fail completely during real droughts like the one in 1980, when our best corn field made a paltry 10 bushels to the acre.

My father was the hardest-working man I ever knew, but no matter how hard he worked he could not make it rain. I watched the despair in his face, from the helplessness of watching the hard-earned fruits of his labor all too often shrivel up and die. It just didn't seem fair that he could work so hard and receive so little. Why did it never seem to rain enough on our farm in the summer?

Every year, I had heard about the wonderful corn crops in neighboring Iowa, and I assumed it rained all summer long there. When I went to college, I found out that Iowa received almost the exact same amount of rain that we did, no more. I also had

| My grandfather, Rollin Strickler, bottle feeding one of his prized Simmental calves

Introduction

roommates from western Kansas who got better corn yields than we did on half the rainfall we received.

How could this be? Were we just stupid? Were we just poor farmers? I later learned in my travels that Iowa and western Kansas had something that southeast Kansas did not: good soil. What exactly was the difference in the soils? When it rained on those soils, the water soaked in. When it rained on our tight, heavy clays, most of it ran off, especially during the intense thunderstorms that occurred in the summer, when soil moisture was needed most. If you dug a hole in those soils of Iowa or western Kansas, you could find roots up to 6 feet deep. In our tight clay soils, the roots stopped about 4 *inches* deep.

We did not have a deficiency of rainfall; we had soils that could not make good use of the rainfall we did receive.

Watching your family's livelihood burn to a crisp every summer leaves an impression on you. I made it my lifetime goal to learn how to reduce our vulnerability to drought. I was determined to make our miserable soil more like those of Iowa or western Kansas: a soil that would let rainfall actually go into the ground, a soil that would hold moisture between rains, and a soil in which roots could go deep and pull that moisture out. I spent hours in the library and sought out every scrap of information I could find on soil improvement and drought management. I interviewed dozens of survivors of the droughts of the 1930s, the 1950s, the 1980s and, of course, 2011 to 2012 and picked up tips and techniques.

Once I began my professional career as an agronomist, specializing in forages and cover crops, I began to network with like-minded people with varied experiences from all over the world. The Internet broadened my access to information about drought, and many late evenings were devoted to scouring the web for the best ideas for improving drought resilience. I learned techniques to enable pastures to withstand drought and ways to feed livestock and water livestock when the best of management falls short. I learned how to improve soil and make it much more capable of withstanding long periods without rainfall.

Since this information is too important to keep to myself, I have chosen to share it in the form of this book.

MAKING A CHANGE

In the year 2000 I bought a farm and gradually began to put my theories to the test. My new place had been managed in what would be considered typical farming practices for the area: it was conventionally tilled, flood-irrigated, and in a corn-soybean rotation.

I began making changes, installing subsurface drip tape, moving to no-till, integrating cover crops, and then planting perennials. In 2004 I dug a trench in this field to put in a drip tape flush pipe. I observed that the roots of the soybean crop then growing on the field grew to a mere 18-inch depth, completely stopping at the point where they hit the clay subsoil layer. In August of 2015, prepping for a farm field day, I dug a 6-foot-deep pit in the same field, so that tour participants could view the roots of the eastern gamagrass growing on the field. The difference was remarkable. Not only did the roots of the eastern gamagrass extend down to the bottom of the 6-foot pit, but the soil at the bottom of the pit surrounding those roots was moist enough that you could squeeze water out of it, despite going without a rain for more than 2 months.

According to my calculations, at the beginning of this project in 2004, my effective root zone held 3 inches of available moisture (1.5-foot depth times 2 inches per foot of depth). In 2015, after 11 years of management designed to create drought resistance, my now 6-foot-plus root zone held over 12 inches of available moisture. Even if you assume this crop could use as much as 2 inches of moisture per week, 12 **acre-inches** of moisture buys a lot of cushion between rains. My theories, accumulated over three decades of research, had finally been put to the test and had passed with flying colors. I now feel compelled to share these techniques with the world.

So how did I accomplish this feat? In essence, I followed this simple recipe:

1. Increase the infiltration of rainfall into the soil and decrease the runoff.

2. Increase the amount of water that the soil profile can store.

3. Increase the ability of plant roots to extract the water when needed.

Or, more succinctly:

1. Get it in.

2. Keep it in.

3. Get it out.

This sounds pretty simple and, really, the basic principles *are* simple. Like so many other worthwhile endeavors, however, the devil is in the details, and that information is what I now propose to share with you. How you can accomplish these three basic principles is outlined in Part I of this book. Part II deals with keeping livestock fed and watered during a drought. Part III consists of two chapters. The first chapter is a plan for changing our agriculture altogether in the most drought-prone areas (the semi-arid Plains), utilizing many of the emergency practices described in chapter 7 as a planned, every-year agriculture system. The second chapter contains a checklist of activities to be performed prior to, during, and after a drought to minimize its impact.

Enjoy.

On January 1, 2000, I began my first year of farming full of arrogance, determined to show the world how much smarter I could farm than everyone else. That first summer, my farm was struck by one of the worst droughts in the history of my area. My irrigation allotment was cut back from 17 to 5 acre-inches, not enough to salvage a decent crop given the meager rain. Adding insult, the soybeans to which I had planted almost my entire farm were sold for the same price that soybeans enjoyed in 1936.

My inability to control how much moisture I received humbled me, humiliated me, and put me in a financial hole from which I have still not yet fully recovered. But that disaster strengthened my resolve to learn how to change what I could control: the ability of my soil to capture, store, and release the moisture it did receive.

PART I

CREATING A MOISTURE-EFFICIENT SOIL

2

GETTING MORE WATER INTO THE SOIL

On May 5, 2015, a tornado hit my hometown. It wrought terrible destruction, cutting through our local ethanol plant and tearing the grain bins into individual sheets of corrugated metal. Those sheets went spinning through the neighboring feedlot at more than 100 miles per hour and sliced hundreds of animals into bits. The tornado destroyed the home of a dear neighbor and dumped up to 7 inches of rain in just a few minutes, causing massive flooding in the area. My creek-bottom pasture was under 3 feet of water, my fences were washed out, and my cattle went on a three-day walkabout.

Very few soils are in a condition to absorb rainfall that intense. My son and I were tornado chasing, and he had his camera pointed out the window at the funnel cloud to the east. As I passed our farm I saw something that made me hit the brakes. All along our drive that afternoon the road ditches were overflowing — until we came to the ditch at the bottom end of my farm, which was empty. According to the nearest intact rain gauge, we had received 4.5 inches of rain in about 20 minutes, and the entire quantity soaked into my field. All my neighbors are excellent farmers, and I am not trying to portray them as bad examples at all. But the rain was pouring off their fields in small gullies several feet across, and every drop that fell on my field *stayed* on my field.

It was a small ray of personal sunshine in the middle of a huge tragedy. It told me that what I had been trying to accomplish with my soil management had worked. It really doesn't matter how much rain you get if most of it runs off and never enters your soil. Therefore, the very first step in being more moisture-efficient is to make sure the rainfall you do receive actually goes into your soil.

THE TILLAGE MYTH

It seems perfectly logical to assume that tillage would increase the infiltration of water into a soil. I mean, just *look* at a freshly tilled soil surface. It is loose and crumbly and porous. In this case, however, looks are quite deceiving. (See Table 2.1.)

Now to be fair, rainfall does enter a tilled surface faster than an untilled one . . . for the first few minutes of a rain. But once raindrops have done their work of pulverizing the exposed **soil aggregates** into a cake-batter consistency, this finely mucky layer becomes an actual barrier to further water infiltration. It doesn't take long for this to happen, either. During an intense thunderstorm, it may take just a few minutes. The USDA Natural Resources Conservation Service (NRCS) has a traveling demo called the Rainfall Simulator that dramatically illustrates how water fails to penetrate a tilled soil.

How Tillage Affects Rainfall Infiltration

TILLAGE TREATMENT	% GROUNDCOVER	MM WATER/ MINUTE
Moldboard plow	12	0.8
Chisel plow	27	1.3
No-till	48	2.7

Table 2.1   Research shows that tillage retards, rather than speeds up, infiltration.

What Are Soil Aggregates?

Aggregation is the clustering together of individual soil particles (sand, silt, and clay) into a spherical conglomerate, held together by biological glues. An aggregated soil will hold plant-available water within the **pores** inside the aggregates, yet will allow the easy movement of rainfall and oxygen into the soil around the aggregates.

A **water-stable aggregate** is one that will withstand wetting and remain in an aggregated condition. A high percentage of water-stable aggregates is a very desirable condition for plant growth. An easy way to measure water stability of aggregates is the **slake test**. This involves placing a lump of soil onto a screen suspended in a graduated cylinder filled with water. A soil with good water-stable aggregation will remain intact, while a soil with poor aggregate stability will quickly melt away and the soil particles will fall through the screen.

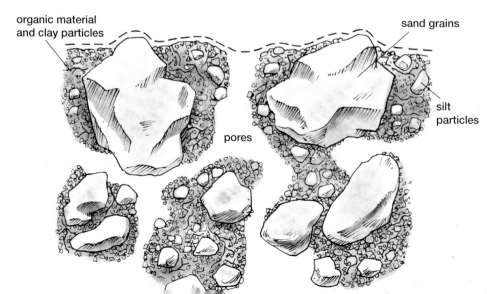

organic material and clay particles

sand grains

silt particles

pores

Short term, the negative consequence of tillage is primarily the exposure of the soil surface to raindrop impact. This impact breaks soil aggregates into individual soil particles, which settle into a dense layer that when dry forms an impermeable crust. The secondary effect is the creation of a **platy layer** at the bottom of the tillage layer, especially with **inversion tillage** like moldboard plowing or disk harrowing. This layer is called a **plowpan**.

To lift a soil up, something has to press down on the layer below it for leverage. The bottom of a plowshare or disk transfers the weight of the implement plus the weight of the soil it is lifting to the layer of soil below it. That pressure flattens those desirable round balls of soil into flat layers that look like plates, with no pore spaces at all. This layer allows water to move through it very, very slowly. For water to enter a pore, air must be able to move out of it. This is very difficult in a platy-structured soil.

In the long term, the biggest negative consequence of tillage is the reduction in soil organic matter. Soil organic matter provides the glues that hold soil aggregates in their desirable round shape, a shape that encourages the free passage of water around them and into the soil profile. Understanding why tillage reduces water infiltration will help clarify the processes and techniques that increase infiltration.

Tillage can create a surface layer of loose soil, but it compresses the layer below the depth of tillage to create a plowpan. This layer has a platy structure that impedes movement of rainwater and oxygen into soil.

HOW WATER TRAVELS THROUGH SOIL

There are two major forces acting on water in soil: 1) gravity; and 2) the attraction of water to the surfaces of soil particles. The attraction of water to soil surfaces is a function of the molecular shape of water, which is slightly bent. This bent shape results in the hydrogen atoms (which are positively charged) being at one end of the molecule and the negatively charged oxygen being at the other. The surfaces of soil particles are negatively charged, so the hydrogen end of the water molecule is attracted to the soil surfaces. The more surface area a soil has, the more attractive force it has for water.

Most of the attraction, or "pull," that soil has for water comes from the clay particles in the soil. Humus is typically such a small portion of the total soil weight that it does not greatly add to the pull that soil has for water compared to the much more abundant clay, though humus has more attraction on a pound-for-pound basis. The pull that soil particles have for water is far stronger than gravity. This is evidenced by the fact that a column of soil placed in water sucks the water up into the column against gravity, like a straw. However, this attractive force extends only a microscopic distance from the soil particle. This means that water even a few millimeters away from direct contact with a soil particle moves under the influence of gravity.

By contrast, water in close contact with soil clings tightly to the soil particle until it is coated with more water than it can hold, a condition we call **saturation**. If soil particles are packed together tightly, water does not move down until each successive layer becomes saturated.

The key to getting water into the soil, then, is to have large pore spaces (**macropores**) open to the soil surface so that water can enter without actually touching soil particles. All of our management to increase infiltration is directly or indirectly related to this goal.

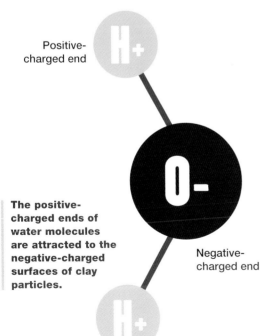

Positive-charged end

The positive-charged ends of water molecules are attracted to the negative-charged surfaces of clay particles.

Negative-charged end

A Film Is Worth a Million Words

If the explanation above is as clear as mud (pun completely intended), please understand that it is difficult to explain in words how water moves into a soil. A visual aid can help. During my teaching aids I liked to show my students the classic Walter Gardner film *Water Movement in Soils*. Fortunately, the film is now available online in both the original and an updated version (see Resources).

MULCH

Perhaps no other practice improves water movement into the soil surface more effectively than creating and maintaining a mulch layer. The primary benefit of a mulch layer is not that it slows the velocity of overland water flow, as is often assumed (though that is important); rather, it is that mulch absorbs the energy of falling raindrops and thereby prevents raindrop impact from destroying soil aggregates. If the aggregates remain intact, the water goes into the soil through the intact large pore spaces, and there is no runoff to slow down.

In my teaching days, I conducted a demo with two areas of soil. Over one area we suspended straw mulch on a frame of chicken wire a few inches above the soil surface. The adjacent area of soil had no protection. We sprayed both areas with a garden hose for several minutes. Even though the straw was not actually touching the soil, there was no

runoff at all in soil under the mulch. The soil surface remained open and loose, while the unprotected area became sealed over and then crusted when it dried. Table 2.2 on this page indicates how surface mulch affects water infiltration.

KEEPING THE RESIDUE WE ALREADY HAVE

So how do we create a mulch? The first and most obvious way is simply to avoid destroying or removing the mulch we have as a byproduct of our current land management, such as crop residue or the stubble of pasture grass.

Keep Plant Residues in the Field

Just as tillage is destructive, so is removal of crop residue from the field.

During the drought of 2012, a huge amount of crop residue was baled and sold for low-grade livestock feed in my area. I had neighbors who thought they had found a miniature gold mine, selling baled corn stalks for $60 a ton. I felt compelled to point out a few items to them. First, the cost of swathing, baling, and moving those bales amounted to around $30 a ton. So, their net above-harvest cost was only about $30. Then I had them figure the value of the fertilizer in those stalks. A ton of corn stalks contains about 20 pounds of nitrogen, about 8 pounds of phosphate, and about 60 pounds of potash. At $0.65 per pound of nitrogen, $0.75 per pound of phosphate, and $0.50 per pound of potash, that fertilizer value amounted to $49 a ton. Essentially, they sold $49 worth of fertilizer for $30.

How Straw Mulch Affects Rainfall Infiltration

STRAW RATE, TONS PER ACRE	INFILTRATION RATE, INCHES OF RAIN PER HOUR
0	0.9
0.5	0.9
1.0	1.7
1.5	2.4
2.5	2.5
3.5	2.5

Table 2.2

The real loss, however, was the value of the stalks as mulch and organic matter. Neighbors who continued on this path for several years started to notice that their dryland fields were dropping in yields quite rapidly. They began to have crop failures while their neighbors were still pulling off average yields. They also had large amounts of potassium deficiency. Potassium deficiency predisposes plants to **stalk rots**, which are fungal diseases that infect the lower stalks. Once the lower stalk is infected, it not only makes it difficult for the plant to take up water and nutrients, resulting in reduced yield, but the structural integrity of the stalk is compromised as well, and the plants begin to **lodge** (fall over) and become difficult to harvest. No one fertilizes with potassium in my area, because our soils tend to be very high in that element. But when crop residue is harvested year after year, the most easily available potassium ions are largely removed from the surfaces of soil colloids and transported away in bales of stalks and straw.

The Amish have a saying that the man who sells hay is slowly selling his farm, and I have seen formerly rich soils become progressively impoverished by too much removal of crop residue. On top of that there is the issue of soil erosion. Removing the protective layer of natural mulch leaves the soil more vulnerable to erosion. I shudder to think of what might happen if cellulosic ethanol ever becomes a viable concern, if crop residue is the preferred feedstock. I am all for green energy, but only when it makes sense in the long run. Being able to fill our gas tank cheaply won't do us much good if we can no longer grow food.

Avoid Overgrazing

Similar to the removal of crop residue on cropland is the overgrazing of pasture lands. It is critical to leave a minimal amount of grass residue on the soil surface to promote water infiltration. The data below illustrate the effects of too much grass removal on water infiltration in a Texas study.

Mineral Nutrient Content of Common Crop Residues

This table shows the high level of nutrients that a field loses when the farmer removes crop residues.

RESIDUE	NITROGEN	P_2O_5	K_2O
Corn stalks	20	7	33
Wheat straw	13	5	20

Table 2.3
Data is approximate, and in pounds per ton. P_2O_5 = phosphorus pentoxide; K_2O = potassium oxide

How Grazing Pressure Affects Infiltration

This study, from Spur, Texas, shows how a 2.5-inch rain infiltrates pasture soils.

PASTURE HISTORY	DEPTH OF INFILTRATION
Heavily grazed	10 inches
Moderately grazed	21 inches

Table 2.4

It may be helpful to point out exactly what "too much" grass removal consists of. In terms of pounds per acre, it is, roughly, grazing so much that less than 2,000 pounds of forage per acre remains. Once residue amounts drop below 1,000 pounds per acre, runoff rates (and evaporation rates, as the

Bringing Mulch In

Of course, we can also simply import organic materials for mulch and apply them to our soil surface. This is obviously much more economical on a garden than it is on fields or pastures, depending on the cost of the material and the cost of hauling and spreading. By all means, if organic mulch can be found at reasonable cost, utilize it.

Be aware that some organic materials that are very low in protein content (like sawdust) can tie up nitrogen upon decay. Soil microbes need nitrogen in the ratio of 1 part nitrogen to about 28 parts of carbon. If the residue is too low in nitrogen, microbes will rob the soil of nitrogen to balance their needs. Wood chips decay much more slowly than sawdust because they have less exposed surface area, and they are much less likely to tie up nitrogen. Some materials, like black walnut and red cedar products, can have a toxic effect on some crops. Always inform yourself about what you apply to your land. In terms of quantity, the general rule is that more organic matter on your soil is better than less.

next chapter will describe) increase even more dramatically. In lieu of clipping, drying, and weighing the forage, however, it is much easier to "eyeball" the pasture. If there is any bare dirt showing (other than in high-traffic areas such as gateways), there has been too much forage removal.

Not only does bare, exposed soil reduce infiltration, but it also means that there is sunlight not being captured by green leaves, and not contributing to pasture productivity. To maximize both rainfall infiltration and pasture productivity, observe the guidelines on page 112 for the minimum residual forage height you should leave remaining after a grazing episode.

NO-TILL COVER-CROP MULCH

On cropland and gardens, we can add to the amount of mulch left by crop residue by growing cover crops in between cash crops. Cover crops can be managed to provide multiple benefits, but perhaps the most beneficial is the addition of surface mulch. As with all mulches, **no-till** is essential to maintaining this benefit.

A cover-crop mulch can dramatically improve infiltration. My very first experience in which I no-tilled (planted a crop without tillage, using a no-till planter) into a cover crop mulch demonstrated this in spectacular fashion. I planted a soybean crop into a killed cover crop of very thick rye. The mulch was so heavy that on 40 acres I flushed 29 hen pheasants out of it while planting (they were looking for a place to nest, and this was by far the best cover around).

During the planting process I had this nagging feeling that there was something I had forgotten. I kept getting out and checking the drill to see what I had omitted. Then it dawned on me: there was no dust. Every time I had planted into tilled ground there was always a cloud of dust that followed the tractor, and a layer of dust on the drill. This time, there was none. I wondered how much longer my bearings and engine would last without all that abrasive soil getting into them (not to mention my lungs) and how much money the lower maintenance alone would save me over the long run.

That summer it rained 11 inches (30 cm) in a two-week period from late July through early August. This is a perfect scenario for growing soybeans, and sure enough, this field produced the highest yield of soybeans the farm had ever grown. But the more astonishing thing about it was revealed later. You see,

this field was shaped like a big bowl, and all the water drained into a small pond. During the two weeks when those 11 inches of rain fell, not once did the level of that pond rise. All the rain went right down into the soil.

Then the rains shut off in mid-August, and no rain fell until the following May 23. I planted wheat into this field following soybean harvest in October, and despite no rain for 10 months, my field raised a whopping 77 bushels of wheat an acre. My neighbor across the field access road, an excellent farmer and winner of many regional yield contests, drilled wheat into his soybean stubble the day before I did; his made 25 bushels an acre. All that excess rainfall that fell on my field during the growth of the soybean crop was stored for use by the wheat crop that followed, instead of running off.

The graphs below illustrate how cover crops improve water infiltration.

How Cover Crops Affect Water Infiltration

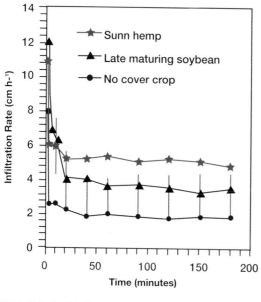

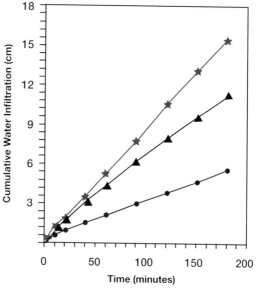

Table 2.5   In this Kansas study, a sunn hemp cover crop roughly tripled the infiltration rate of no-till wheat stubble without a cover crop.

MANURE

Animal manure can be a huge benefit to water infiltration, through several mechanisms:

1. The typical fibrous nature of manure acts to absorb raindrop impact just the way plant residue does.

2. The decay of manure produces organic glues that promote aggregation.

3. Manure is a food source for earthworms and dung beetles that create large macropores that allow water to enter the soil surface and rapidly move into deeper layers. The burrows left by these little animals can allow the movement of huge volumes of water into the soil profile in a very short period of time. In a well-managed pasture with abundant earthworms and dung beetles, cowpies disappear in a few days, leaving behind just the surface-dried crust of the cowpie and a vast network of tunnels below it.

How Manure Affects Rainwater Infiltration

AMOUNT OF MANURE	PERCENTAGE OF RAINFALL THAT INFILTRATES SOIL
None	72%
8 tons manure per acre	81%
16 tons manure per acre	90%

Table 2.6

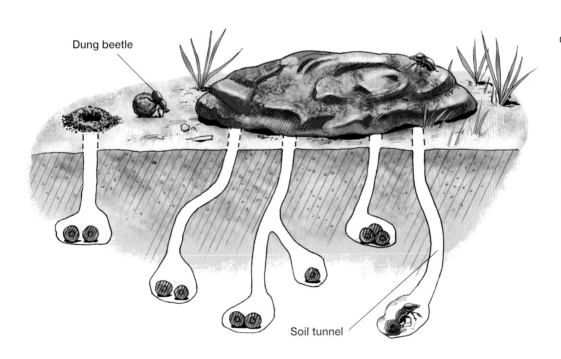

Dung beetle

Soil tunnel

A cowpie acts as a food source for many soil-tunneling animals, such as earthworms and dung beetles, which create macropores that enhance water infiltration.

LET ANIMALS SPREAD IT THEMSELVES

Most research has been conducted on manure that has been mechanically spread. I contend that a greater benefit comes from letting the animals spread the manure themselves. Not only is this more energetically and economically efficient, but the manure is deposited in the form most preferred by earthworms and dung beetles as well. I have never seen dung beetles attracted to mechanically spread manure: it is no longer fresh, and dung beetles prefer fresh manure. Earthworms likewise prefer the well-insulated thermal environment below a cowpie to small fragments of manure spread across a landscape.

It is easier to move the animals and the feed to wherever the manure should be deposited than to move the animals and feed to a lot, then scrape up the manure, haul it, and spread it. Techniques of self-spread manure, such as spaced bale feeding and bale grazing, are designed to accomplish this task. The difference between these two techniques is subtle.

Spaced bale feeding means bales are set out on a grid pattern, roughly 30 feet by 30 feet apart and enclosed by an electric fence, with one side of the fence being portable polywire on a reel. Once the feeding period begins, you expose a few bales — preferably one day of feed at a time — by rolling up the polywire and moving it past the first row of bales. Using two polyreels and "leapfrogging" them makes it easier to keep the animals where you need them. A 30- by 30-foot spacing results in about 50 bales on an acre.

Bale grazing involves unrolling the bales, with the unrolled strips being about 20 feet apart. Again, one day's worth of feed is exposed at a time. If each bale unrolls out to 200 feet in length (this varies tremendously) on a 20-foot spacing, this makes for about 10 bales per acre.

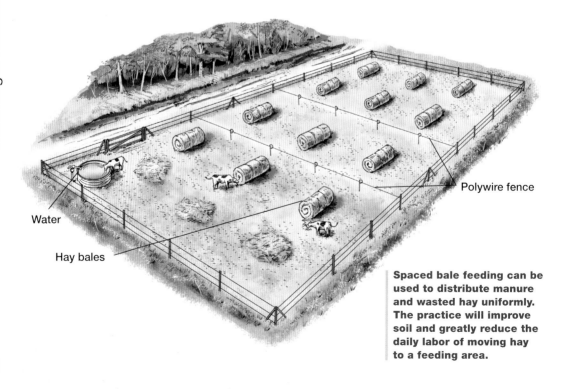

Water

Polywire fence

Hay bales

Spaced bale feeding can be used to distribute manure and wasted hay uniformly. The practice will improve soil and greatly reduce the daily labor of moving hay to a feeding area.

Bale grazing,
before and after

Do the Math

The uneaten hay and manure and urine combine to make a very rapid improvement in soil organic matter and fertility. Each ton of dry hay fed results in the production of about 2 tons of wet manure (25 percent dry matter). A ton of alfalfa hay contains about 60 pounds of nitrogen, 12 pounds of phosphate, and 60 pounds of potash. Since about 75 percent of the nitrogen and 80 percent of the phosphorus and potassium remain in the manure or urine, animals bale-grazing ten 1-ton bales per acre deposit about 20 tons of fresh manure, 450 pounds of nitrogen, about 100 pounds of phosphate, and 480 pounds of potash on an acre.

Spreading the Wealth

The distribution of that organic matter and those nutrients is not necessarily uniform, and since they are deposited during a time when earthworms and dung beetles are inactive, harrowing or dragging may improve uniformity of nutrients across the feeding area. However, harrowing also spreads any parasites and disease organisms over a larger area, and reduces the value of the manure to earthworms. I prefer to just let the worms do the work, although I must admit that harrowing cowpies behind an ATV is a lot of fun.

Spaced bale feeding deposits five times the amount of organic matter and mineral nutrients as bale grazing and is the single fastest technique I know to raise soil organic matter on a field scale. An acre of legume-based pasture so treated should never again need fertilizer in a lifetime. I somewhat prefer the bale grazing, however, because I can treat more acres in this manner given the same number of bales. As old as I am getting, I would prefer to have as many of my acres

exposed to one of these two methods as soon as possible.

MYCORRHIZAL FUNGI

Mycorrhizal fungi are symbiotic organisms that colonize plant roots and send out tiny filaments called **hyphae** that act like root hair extensions, aiding in the uptake of water and nutrients. They also secrete a lubricant substance, called **glomalin**, that acts like a glue to improve soil structure. Glomalin is also very persistent in soil, lasting as long as 42 years. The dramatic improvement in soil aggregation after mycorrhizal colonization results in a significant increase in water-stable aggregates, with large pore spaces that enhance water infiltration. (Table 3.16 on page 61 illustrates the effect of glomalin on water-stable aggregates.)

Mycorrhizal fungi need a living root system in order to survive. Unfortunately, the worldwide conversion of natural ecosystems based on perennial plants to systems that use annual crop plants has resulted in long fallow periods between crops, which has caused starvation of mycorrhizal fungi. Now almost all cropland is devoid of these organisms, as well as much pastureland that was formerly cropland.

The good news is that now there is commercial inoculant available for these organisms. I have personally used a mycorrhizal inoculant called MycoApply and have found it to be very successful at improving soil structure. It has given me a nice boost in plant production, particularly on my more difficult soils, such as salty spots, heavy clay spots, and eroded areas. The results are absolutely dramatic on reseeded warm-season native grasses, like eastern gamagrass or big bluestem. MycoAppy is available in many formulations (granular, powder, and liquid) and is applied at a rate to supply roughly 90,000 propagules (spores) per acre.

The spores are the "seeds" of the fungi, and the hyphae are the "plants." The spores are quite tough: they are not harmed by ultraviolet light, by temperatures up to 140°F (60°C), or by drying out, and can survive in the bag, on the seed, or in the soil without a host for up to two years. Once the spores germinate and become hyphae, on the other hand, they need a live root as a host, and will starve within a few weeks if the host plants are killed.

The best way to apply these inoculants is with seed during planting, so the spores can contact roots. They can be broadcast on top of the soil, but it takes a good soaking rain to move the spores into the root zone. See Resources for more information on obtaining mycorrhizal fungi inoculant.

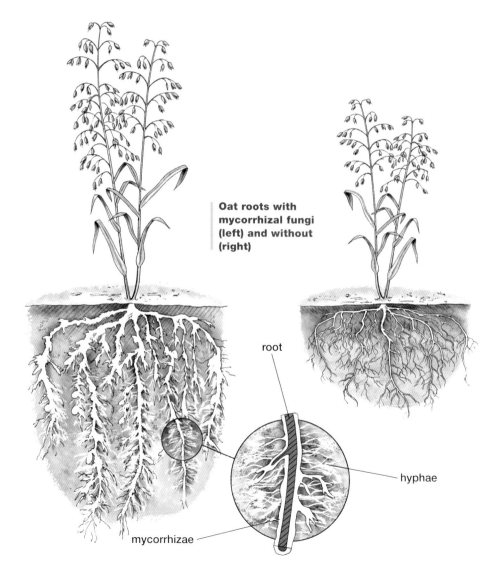

Oat roots with mycorrhizal fungi (left) and without (right)

root

hyphae

mycorrhizae

EARTHWORMS

Earthworm burrows can become easy passageways for water into the soil. Not only is the opening of the burrow an access point for water, but the slime secreted by the worms as they move about also improves soil aggregation along the burrow lining. The combined effect is that water moves well into the tunnel and also moves easily out of the tunnel and into the soil.

A few simple practices can improve worm population and activity: no-till, maintenance of surface cover, cover crops, and manure applications. Also, it's best to avoid soil-applied insecticides from the organophosphate or carbamate groups. Vegetation that includes legumes is tremendously beneficial in increasing earthworm populations.

How Earthworms Affect Soil Moisture

This study took place in New Zealand.

DATE	MOISTURE CONTENT (PER CUBIC METER OF SOIL)	
	PASTURE WITHOUT EARTHWORMS	PASTURE WITH EARTHWORMS
June	7.9 cm	8.5 cm
August	6.6 cm	8.4 cm
November	2.5 cm	3.4 cm
February	5.6 cm	6.4 cm

Table 2.7

How Tillage Affects Earthworm Populations

TILLAGE SYSTEM	EARTHWORMS PER CUBIC METER OF SOIL
Chisel plow	67
Ridge till	178
No-till	211

Table 2.8

How Diet Affects Weight of Nightcrawlers (*Lumbricus terrestris*)

FOOD ITEM	WEIGHT CHANGE, PERCENT
No food	-11%
Bromegrass leaves	-11%
Corn leaves	+3%
Red clover leaves	+19%
Alfalfa leaves	+35%

Table 2.9

How Cover Crops Affect Earthworm Populations

This data describes fields in a sorghum-wheat rotation.

COVER CROP TREATMENT	EARTHWORMS PER CUBIC METER OF SOIL
None	2.8
Forage soybean	10.8
Sunn hemp	18.5

Table 2.10

How Cropping Systems Affect Earthworm Populations

CROP	TILLAGE SYSTEM	EARTHWORMS PER CUBIC METER OF SOIL
Continuous corn	Plow	10
Continuous corn	No-till	20
Continuous soybeans	Plow	60
Continuous soybeans	No-till	140
Bluegrass/clover	none	400
Dairy pasture + light manure	none	340
Dairy pasture + heavy manure	none	1,300

Table 2.11

Earthworms: Too Valuable for Fish Bait

To summarize, worms are good. To get more worms:

- No-tillage is better than tillage.

- Legume crops (alfalfa, clover, soybeans) are better than grass crops (corn, bromegrass).

- Perennial crops (pasture) are better than annual crops (corn and soybeans).

- Manure applications are good. Insecticides are not.

- High-protein cover crops are better than no cover crops.

A forage radish leaves behind a significant root channel.

ROOT CHANNELS

Old, decayed root channels act as passageways for water into the soil. For these roots to benefit water infiltration, no-tillage is essential.

Even a little tillage is harmful. The decayed root must form a continuous open channel from the soil surface into the soil in order to be effective. Have you ever been snorkeling? There is a big difference between the top of the snorkel being an inch above the surface of the water versus an inch below the surface. Just as the water cuts off the supply of air to the snorkel if the top of it does not reach the water surface, soil can cut off the path for easy entry of water into the soil if the root channel is severed by tillage and no longer reaches to the soil surface. Even a shallow tillage operation cuts off the old roots and covers the channels with soil, disrupting the connection to the surface.

While the decayed roots of old crops offer some benefit to improved water infiltration, the best value by far comes from creating additional root channels with cover crops selected for the purpose. For example, a corn crop planted at 30,000 plants per acre gives 30,000 root channels per acre; a soybean crop may have 150,000. Compare those to a rye cover crop, which can provide more than 2 million root channels per acre. Tossing some daikon-type radish into the cover crop mix can add another 100,000 very large and deep root channels to the soil per acre.

For the maximum benefit, blend fibrous-rooted cover crops with deep taprooted crops. Fibrous-rooted cover crops include annual ryegrass, rye, barley, buckwheat, and phacelia. Taprooted cover crops include sweetclover, brassicas (such as turnip and radish), okra, and sunflower.

LANDFORMING

A highly effective, though expensive, way to improve water infiltration is to shape the soil surface to prevent runoff. Methods include terraces (including contour tillage), vertical mulches, and retention dams.

TERRACES

Contour tillage involves making corrugations or ridges along contour lines so that each little ridge acts like a miniature dam to hold water. Terraces come in many forms, from broadbase terraces, designed to be driven over, to level-bench terraces, created by

A tile inlet is used to dispose of "excess" water that ponds in terrace channels. This occurs when rainfall runs off soil instead of infiltrating it. A better system is to capture as much water as possible and store it in the soil. Improved infiltration, better internal drainage, and more open pore spaces will reduce the risk of drowning out crops.

Terracing in perhaps its finest form may be seen at the ancient Inca farms of Ollantaytambo, found in the Sacred Valley of modern-day Peru. This is a form of agriculture that persisted at a high level of productivity for centuries.

using rock walls to turn a slope into multiple sections of level ground. An example is the ancient region of Ollantaytambo in what is now Peru; see lower photo on page 29.

Each method can be effective at reducing runoff, but neither is a complete solution to prevent runoff. If they were, we would not need tile outlets to conduct away "excess" water as shown in the upper photo on page 29.

Swales are a form of terrace set on a contour. A swale differs from the typical American terrace, which is designed to gently remove runoff water from a field without causing erosion and has a gentle slope (usually 1 percent) leading to a waterway or tile inlet. By contrast, a swale is created to be level so that runoff is captured and kept on the field as long as possible. Swales are often planted with a flood-tolerant vegetation, such as pecan trees, eastern gamagrass, reed canarygrass, or willow trees.

If swales are placed on the side of the field that is upwind during prevailing summer winds, they can both slow and humidify the air passing over the field in summer. Typical American terraces are designed to treat water like a problem; swales are designed to treat it like an asset.

Swale

Wood chip mulch to promote infiltration

A swale is designed to temporarily impound water and allow it to soak into the ground instead of running off.

VERTICAL MULCHES

To build a **vertical mulch**, you cut small trenches along contour lines, throwing the soil from the trench to the downhill side to form a mini-dam, then filling the trench with coarse organic material, such as manure, straw, wood chips, or shredded paper. This technique, though expensive, is extremely effective at promoting infiltration, especially in very heavy clay soils.

A friend in Texas farms on blackland soils, which are very heavy clays with high shrink-swell potential. When I told him about this vertical mulch idea, he thought it made complete sense. Then he gave it a bit more thought and asked why we need to spend the money to make a trench when his entire farm develops 2-inch-wide cracks in the ground as soon as it gets dry. Why not just spread the coarse organic material across the landscape when the soil is already cracked, and let the next rain push it into the cracks? I could not disagree!

Read more about vertical mulches on page 36.

Vertical Mulch

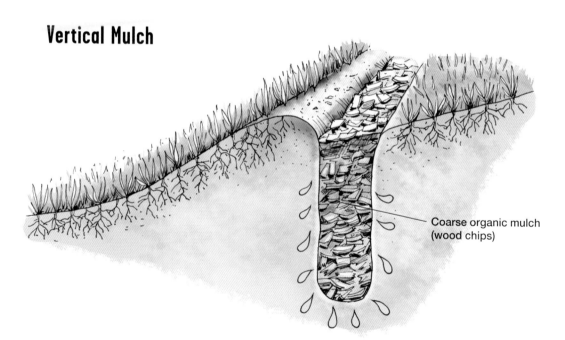

Coarse organic mulch (wood chips)

A vertical mulch can be an effective tool for encouraging infiltration and aeration of soils, particularly heavy clays.

Getting More Water into the Soil

RETENTION DAMS

To create a **retention dam**, first locate the paths where water tends to accumulate during runoff events. Then build a series of small dams to trap the water and hold it as high as possible on the landscape.

Retention dams should be no more than a couple of feet high; otherwise, the force of the trapped water puts too much pressure on the dam. Cover the spot immediately below each dam in rock to keep the force of the falling water from scouring out a gully there. The water from each pool can be tapped for livestock water or irrigation when needed.

Retention Dam

A seasonal watercourse before and after the installation of retention mini-dams

before

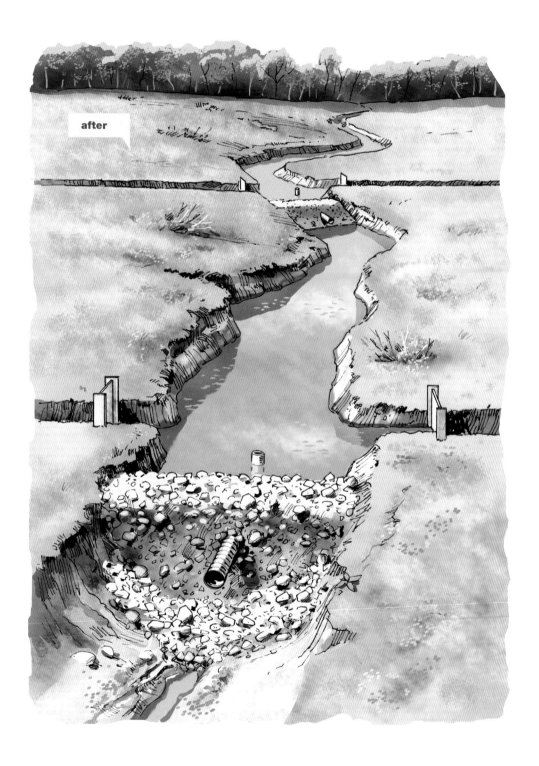

after

KEYLINE PLANNING

It is said that great minds think alike. While throughout the ages innovators in the American Southwest have been creating mini-dams with rocks to make the desert more productive, a man on the opposite side of the planet had a similar idea, but on a larger scale. P. A. Yeomans was a former gold-mining engineer turned cattle rancher in Australia. He adapted techniques he learned while providing water for mining operations, using them to improve soil moisture and live-stock water on his ranch.

He observed that even in arid areas there is often a surplus amount of water in val-leys, while the adjacent ridgetops are quite parched. He created a system that would move water out of the valleys and onto the ridgetops to the benefit of both areas. His ideas, first described in his 1954 book *The Keyline Plan*, involved identifying **keypoints** in the landscape. A keypoint is a location at the top of a valley (see diagram) where a steep slope begins to flatten out. This is the highest point on the landscape where water can be economically impounded.

Yeomans then identified the **keyline**, which is a contour line on the landscape that is on the same elevation as the keypoint. He then created a furrow on this contour line to take overflow from the pond at the keypoint out to the ridgetops, where another impound-ment can be created to hold a reserve of water on top of the ridge, which he called a saddle pond. Since the saddle pond sits on the ridgetop, the water in the saddle ponds can be distributed by gravity to any point on the ridge that is lower in elevation, to be used for either irrigation or livestock water.

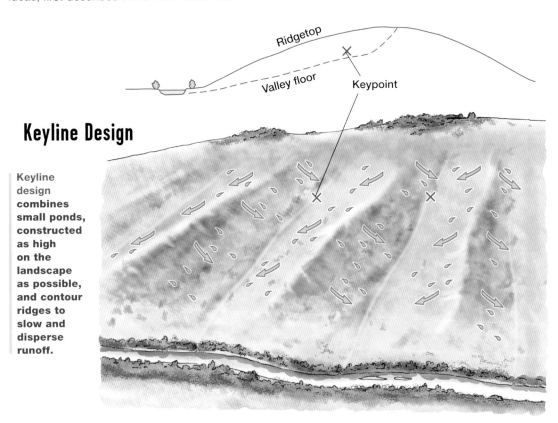

Keyline Design

Keyline design **combines small ponds, constructed as high on the landscape as possible, and contour ridges to slow and disperse runoff.**

Finally, he created a series of furrows parallel to the keyline with an implement similar to a subsoiler (see page 68), which became referred to as a Yeomans plow. If you study the topography carefully, lines parallel to the keyline and below it will have a slight slope downward to the ridgetops, and the net effect is to move runoff water toward the ridgetops. Yeomans claimed that the use of what he named the keyline system would increase topsoil depth, thanks to the use of the Yeoman plow to aerate the root zone.

Despite the existence of keyline systems for more than 70 years around the world, there has been a conspicuous lack of research on the topic. The claims of improved topsoil depth may be unsubstantiated, but the effect of landforming to better distribute water and hold it high on the landscape is clearly beneficial. Personally, I am dubious about the benefits of the Yeoman plow (see discussion on subsoiling in chapter 4), but I am a firm believer in the other components of the system. I think keyline planning is particularly well suited to pastures, since much of the sloping, gullied land that benefits most from keyline management is (or should be) in pasture.

After reading Yeomans's book, you will look at landscapes entirely differently. Instead of seeing poorly productive, gullied pastures, you will begin viewing these areas as potential candidates for installation of keyline systems. I am far from an expert on keyline planning. If this is a topic that interests you, and you desire further information or assistance, I recommend you contact Abe Collins (see Resources, Collins Grazing).

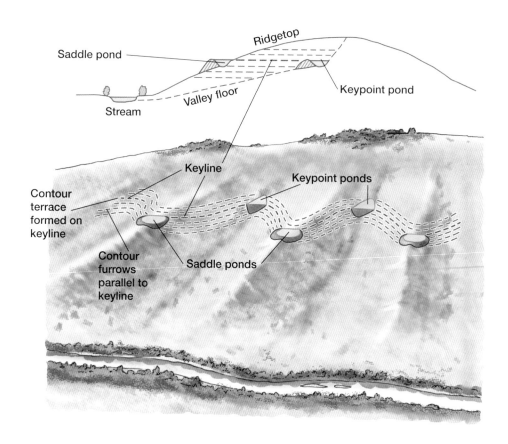

K-LINE IRRIGATION SYSTEMS

Once you have created water impoundments high on the landscape with keyline planning or mini-dams, you can put this water to beneficial use for livestock drinking water or irrigation. Since these small impoundments seldom hold a large amount of water in any one place, they rarely have enough water to justify installation of a traditional fixed-in-place irrigation system such as a center pivot. The ideal irrigation system to make use of these small water reservoirs would be both inexpensive and portable, so that one system can be moved from one impoundment to the next as the water is used up.

Such a system exists, and it is called K-line irrigation. K-line is the brand name for a system of small sprinklers nestled in tough plastic tubs that are connected in a series via a water supply hose, and this system is towable from place to place by any small vehicle such as an all-terrain vehicle. K-lines require only a small pump to operate, and only a small amount of water. They are towed to a new location on a frequent basis in a zig-zag pattern, and can be simply picked up and moved from pasture to pasture.

K-lines are best suited to pasture, since they can be towed only in very short vegetation as might occur after a grazing event. This is a great system for areas that need only occasional irrigation during drought periods.

A VERTICAL MULCHING MACHINE?

Creating vertical mulches, one at a time, is a laborious job. What if this process were mechanized? Soil scientists at the Dale Bumpers Research Center in Booneville, Arkansas, have built just such a machine, which they call the subsurfer. Here's an overview.

K-line irrigation pods in action. These inexpensive, portable irrigation units are ideal for utilizing small sources of water, like small pasture ponds, for pasture irrigation.

- The subsurfer was designed to reduce odors, nutrient runoff, and ammonia volatilization from poultry manure by placing the manure in a narrow slit in the soil instead of broadcasting it on top of the soil surface.

- It is essentially a walking floor trailer that moves manure to the front and drops it into modified no-till planter units with a very large, rectangular seed tube, spaced 15 inches apart.

- If this machine is operated on the contour to the slope, each little furrow is a vertical mulch trench, and an easy entry point for runoff water to infiltrate the soil.

- While it was designed to use poultry litter, it could also easily make vertical mulches with any fine organic residue, such as compost.

As of this writing, the machine is not yet in commercial development and currently exists only in prototype form. Research results from use of this machine, compared to surface-applied poultry manure, indicate that in addition to providing water management benefits, the reduction in nitrogen losses from runoff and volatilization nearly doubles the nitrogen use efficiency of the manure.

CHAPTER SUMMARY

The first step in being more efficient with rainfall is to get all the precipitation you have received into the soil. A number of techniques can effectively accomplish this task, including the following:

- Eliminate tillage in order to retain surface residue from previous crops, and leave root channels intact

- Create additional mulch with cover crops

- Leave adequate residue for complete soil coverage during grazing

- Apply manure

- Manage for earthworms

- Inoculate with mycorrhizal fungi

- Use cover-crop blends that contain both coarse taproots and fine fibrous roots

- Build terraces and swales

- Create vertical mulches

- Use keyline planning with small retention dams to impound water as high as possible on the landscape

3

KEEPING WATER IN THE SOIL LONGER

During my childhood, it was very often too muddy to get into the field for weeks at a time. When we finally did, it seemed to go from mud to brick almost overnight. The moisture that seemed so abundant just a couple of weeks earlier was soon gone, and crops began to suffer unless it rained shortly thereafter.

Why were we losing moisture so fast? I didn't know it at the time, but the rapid loss of moisture from our soil was due to a century of soil mismanagement, just like our lack of infiltration. The previous chapter focused on getting water into the soil. This chapter focuses on retaining it in the soil until it can be used by the plants. If we understand the mechanisms by which soil moisture loss occurs, we can reduce the speed at which it occurs.

HOLDING WATER IN THE SOIL

There are three ways that water is lost from a profile once it has entered the soil:

1. Transpiration (use by plants)
2. Leaching below the root zone due to inadequate water-holding capacity
3. Evaporation from the soil surface

We will describe how to reduce water loss that occurs from each.

CONTROL TRANSPIRATION BY CONTROLLING WEEDS

Transpiration is the loss of water vapor from plant leaves during the photosynthesis process. Since our goal is to grow plants, and transpiration is necessary to grow plants, it can be seen as a necessary and desirable use of water, as long as the plants are desirable. If the plants are weeds, then managing or eliminating those weeds can conserve water for more desired plants.

How Much Water Plants Use

The amount of water used by plants is often quite surprising. This chart compares **transpiration ratios**, the weight of water needed to produce the same unit weight of several different crops and weeds.

PLANT	POUNDS OF WATER USED TO PRODUCE 1 POUND OF PLANT
Sorghum	305
Corn	350
Wheat	557
Alfalfa	720
Soybeans	738
Kochia	260
Pigweed	305
Russian thistle	314
Smartweed	678

Table 3.1

As shown above in table 3.1, preventing a pound of smartweed growth can save about 678 pounds — more than 80 gallons of water. This same amount of water can grow about a pound of crop biomass, or about a pound of pasture grass. Also note that some plants are much more efficient than others. We will revisit this later.

Timing Weed Control in Cropland

Most people try to do a good job of weed control in cropland, but the timing is often too late. Since the development of Roundup Ready crops, it has been increasingly commonplace to "wait until they are all up" and then try to get by with spraying only once. The result is that the weeds are allowed to compete with the crop for several weeks prior to being killed (at least they were killed in the era before all our weeds became resistant to Roundup). This has a dramatic though often invisible effect on the crop.

Light reflected off of weed growth has a different wavelength than the light reflected off of bare ground or crop residue. This particular wavelength of light triggers a response in the crop that causes it to produce more of a plant hormone called **auxin**, and grow taller. At first glance the effect appears beneficial, but the hidden price is that this increased height comes with the expense of drastically reduced root growth. Therefore, the crop becomes much more susceptible to dry conditions later.

We have tremendous tools for controlling weeds in crops; using them early costs no more than using them late and can create a much more drought-tolerant crop. Don't fall into the trap of "revenge killing" of weeds after they have already cost yield.

Rethinking Pasture Weeds

In pastures, rampant weedy growth is probably the norm rather than the exception. Controlling these weeds can have a major impact on the conversion of soil moisture into saleable product — but finding ways to utilize the weeds, rather than simply eliminating them, is usually far cheaper and more effective.

It is essential to recognize that many plants we consider weeds may actually have benefits we aren't aware of, so be cautious about large-scale herbicide applications. Broadcast herbicide applications in pastures can kill a number of valuable **forbs** and legumes and actually reduce pasture productivity. It is impossible to be a good pasture manager without knowing what plant species are in the pasture. If you cannot identify all your pasture plants, get the resources you need to educate yourself. The weed control method you choose may be detrimental to desirable pasture species if you don't know what they are.

There may be other detrimental effects as well. When I was a kid, my father sent me out with a hand sprayer full of **2,4-D** and told me to kill all the milkweed plants I could find, because they are poisonous to cattle. I spent years unwittingly destroying the sole larval host plant of the increasingly rare monarch butterfly. I also found out that while milkweed is poisonous to cattle, cattle never actually eat it, and it doesn't seem to reduce grass production at all.

Likewise, I spent a rather tidy sum of money on herbicides to eliminate velvetleaf from my pastures, only to discover that when I converted to **management-intensive grazing,** the cows ate the plant. For some reason that I may never understand, when I move cattle daily, they eat velvetleaf. When I give cows a multi-day pasture allotment, they don't eat velvetleaf. It is not a matter of the cattle starving to the point where they will eat anything, as I try to leave a very tall residual upon exiting a paddock so there are a lot of other plants available to eat.

I don't know how the cows know they are going to be in a paddock only for a day and not several days. But I do know that with daily paddock shifts, cattle will eat many plants that are considered weeds, including giant ragweed, velvetleaf, marestail, and many others they do not bother even nibbling when continuous grazing. I don't understand it, but I have seen it happen often enough to confirm it. I used to laugh at the guys who told me that cows eat all the weeds with management-intensive grazing, and I assumed they were starving their cows down to where they'd eat anything. Now I *am* one of those guys, and people look at me the same way I used to look at those I used to ridicule.

When livestock move to a fresh paddock every day, they consume virtually all of the plants to some degree. This interesting change in animal behavior, which occurs with daily rotational grazing, transforms many plants we consider weeds into yet another forage.

Turning Waste into Taste

When I was a kid, we had a small feed-lot on our farm where we fattened beef calves. One day, six wild hogs got in our feedlot and just decided to stay. We didn't feed them at all. They survived solely by eating the grain-rich manure patties of the steers, and they got quite fat. We rounded them up one day and butchered them, and bought an old used freezer to store all the additional meat.

I have never since tasted pork that flavorful. It was not the "other white meat" you now buy in the store. It was dark red from the iron content of the meat, juicy, and wonderful. It turns out that ruminant manure is very high in many of the essential nutrients that swine need, including B-12 and iron, two nutrients we must

supplement for swine receiving a grain-only diet. I believe the primary contributor to the wonderful flavor of pork raised in a pasture setting along with cattle is their consumption of cattle manure, which is also a big part of the meat's higher nutritional value. Some people think that eating manure-fed pork sounds disgusting, but the manure they eat never gets any closer to the meat of the pig than the manure the pig produces itself.

I can personally attest to the added flavor. I still remember the day we pulled out the last package of that ground pork from the freezer, and the sadness the entire family felt when it was gone. Turn waste into taste. The water used to grow the plants that the cattle manure came from can be used to feed two sets of animals. Now that is water efficiency!

Enlist Other Grazers to Help

Another method to convert "weeds" into feed is to add additional livestock species. Goats eat many plants that cattle do not, as do sheep and even horses. I have found that horses do not eat Illinois bundleflower and cattle love it; conversely, cattle do not like yellow fox-tail, while horses relish it. Goats love smooth sumac and rough dogwood, and sheep love ironweed, but cattle ignore all of these.

There are many benefits to having multiple livestock species in the same pasture in addition to the consumption of more weed varieties. Have you ever noticed how cattle avoid the forage just around a cowpie? That is

a survival mechanism to keep down parasite populations, since intestinal parasites hatch in cowpies and crawl up adjacent forage. Sheep and horses will eat that forage right to the cowpie edge, however. Parasites of cattle intestines do not survive inside a horse, sheep, or goat, nor do the parasites of those species survive inside cattle.

An even better synergy develops if you add nonruminants to the mix. Pastured swine or poultry actually can use the cowpie itself as a nutrient source, and in doing so consume the parasites and spread the fertility around.

Try a Weed Wiper

Another method of low-cost, selective weed control is to use a **weed wiper**, such as that manufactured by GrassWorks (see Resources). This works very well with management-intensive grazing. First, you graze a paddock off, using enough grazing pressure so that all edible plants are eaten down to a desired stubble height but the weeds are left at full height. Then, you pull the weed wiper filled with a powerful concentration of herbicide across the paddock. Since the wiper applies herbicide just to the uneaten weeds, only small amounts of herbicide are used and susceptible plants are spared. This method is both cheap and environmentally friendly yet still allows us the use of powerful herbicides in our toolbox.

You can control broomsedge (a weedy grass) in grass pastures by wiping the standing plants with glyphosate (which would kill pasture grass if broadcast sprayed). You can kill thistles in clover pastures by wiping them with aminopyralid (Milestone), which would kill clover if broadcast sprayed.

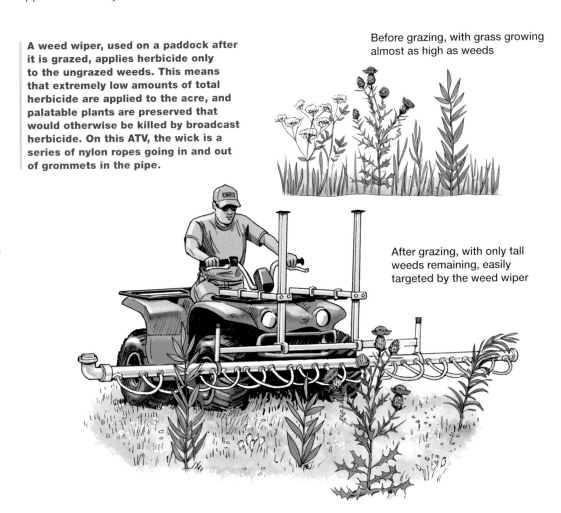

A weed wiper, used on a paddock after it is grazed, applies herbicide only to the ungrazed weeds. This means that extremely low amounts of total herbicide are applied to the acre, and palatable plants are preserved that would otherwise be killed by broadcast herbicide. On this ATV, the wick is a series of nylon ropes going in and out of grommets in the pipe.

Before grazing, with grass growing almost as high as weeds

After grazing, with only tall weeds remaining, easily targeted by the weed wiper

A weed wiper can selectively remove weeds in pastures at low cost. The working piece is the wick attached to the reservoir (PVC pipe). On this pull-behind type, the wick is fabric wrapped around the PVC pipe reservoir.

To the left is a closeup view of the wicking fabric used in one brand of weed wiper. Above is a closeup of the fill cap of the herbicide reservoir.

Teach Your Livestock to Like Weeds

Some folks have even developed methods of training animals to eat weeds. Dr. Fred Provenza, professor emeritus at Utah State University, has made a career out of determining why animals eat some plants and not others with similar nutrient content. His work is quite fascinating, explaining how secondary compounds, such as terpenes, alkaloids, tannins, and saponins, interact to influence plant palatability.

For example, if a cow eats a plant high in tannin, she can tolerate higher levels of alkaloids in her diet. Interseeding high-tannin plants (such as lespedeza or trefoil) into a pasture that contains high-alkaloid plants (such as wild reed canarygrass or endophyte-infected tall fescue) may encourage more consumption of those weeds. Kathy Voth, a disciple of Dr. Provenza, has written a book, *Cows Eat Weeds* (see Resources), on this subject.

Another method I have heard of to encourage weed consumption is to fill the weed wiper with molasses instead of herbicide, to make the weeds tasty to the animals. This must be done just prior to moving the animals out of the paddock. The animals get a sugar buzz (I've always thought cattle and children have a lot in common) and begin to associate it with the taste of the plant itself. Eventually, they eat the plant without the molasses. I would hesitate to use this method if you have poisonous plants, however. Again, I stress the importance of knowing your pasture species.

In a rotational grazing system that features daily moves, animals will eat weeds, as demonstrated by this calf with a mouthful of cattails.

Management through Mowing

A non-chemical way to manage unpalatable pasture weeds is simply to mow a paddock after the animals have grazed off what they want. This usually doesn't kill the weeds, but it often keeps them from going to seed and makes them much less competitive. This method is expensive, though, and does not put the weeds to beneficial use.

REDUCING TRANSPIRATION BY MAXIMIZING CO_2

It is a little-explored fact that plants become more water-efficient as the amount of carbon dioxide (CO_2) available to them increases. This is because plants obtain CO_2 for photosynthesis by opening little pores in the leaf surface called **stomata**. As long as these pores are open, the plant loses water vapor through them. The exiting water vapor from the leaf creates tension in the plant's vascular system that pulls water from the soil into the roots.

In times of water shortage, plants have the ability to get all the CO_2 needed for photosynthesis without the stomata being wide open. This can greatly reduce water use. By increasing the CO_2 in the crop canopy, you can help the plants become more water-efficient because they can tighten the stomatal openings, reducing water loss, and still obtain all the CO_2 they need.

This effect is far more pronounced with plants having a C_3 method of photosynthesis (such as barley, wheat, or soybeans) than plants with a C_4 pathway (such as corn). See Table 3.2.

How CO_2 Enrichment Affects Barley Yield

This Australian study showed the effectiveness of CO_2 enrichment under greenhouse conditions.

	CONTROL	CO_2 ENRICHED
Grain yield (g/m²)	277	471
Straw yield (g/m²)	511	967

Table 3.2

HOW TO MANIPULATE CO_2 LEVELS IN THE CANOPY

- **Use no-till planting methods.** Preplant tillage results in an immediate flush of CO_2 from microbial activity within just a few days. The CO_2 exits the soil before a crop canopy is there to intercept and use it. No-till allows natural decay to release CO_2 in concert with canopy development.

- **Keep high levels of decaying organic matter on the soil surface.** Using cover crops can keep surface decay processes active, and no-till keeps organic matter on top along with existing crop residue. In pastures, feeding hay or topdressing with compost can add surface organic matter.

- **Add surface applications of finely ground limestone if soil is too acidic.** Limestone reacts with acid to release water, calcium, and carbon dioxide. A ton of pure calcium carbonate lime, when fully reacted, releases 1,200 pounds of CO_2. Lime on the surface breaks down much more slowly than lime that has been incorporated with tillage, and it releases CO_2 more slowly, in synchrony with the crop. However, this may be slower than desired for the purpose of adjusting soil pH. Monitor soil pH and be proactive with smaller but more frequent applications of lime.

- **Any measure designed to improve soil oxygen levels** tends to improve the level of CO_2 in the canopy, generally speaking. Carbon dioxide is detrimental in the soil atmosphere because it displaces soil oxygen, but it is beneficial in the crop canopy. Therefore, getting CO_2 out of the soil and into the canopy is simultaneously beneficial to both root function (by displacing soil CO_2 with oxygen) and crop photosynthesis and water-use efficiency. Nightcrawler activity is especially useful for pumping air in and out of soil in morning and evening.

REDUCING EVAPORATION

Evaporation from the soil surface is increased by three main factors: wind speed at the soil surface, sunlight hitting the soil surface, and soil temperature. You can enlist plants themselves — trees, standing crops, and cover crops, for example — to protect the soil from excessive evaporation in a variety of ways.

Reducing Wind Speed

Windbreaks or shelterbreaks of trees reduce wind speed at the soil surface (standing corn and even tall crop residue can do the same, which we'll discuss). Research has proven time and again that woody windbreaks pay for the land they occupy with increased yields from an area 10 times their height. Sadly, every time crop prices get high, millions of tree windbreaks are bulldozed to gain a few extra acres at great cost, and the truth is that while there is more acreage in production, the net effect is a reduced total yield from those fields.

I have neighbors who spent $20,000 to bulldoze a half-mile windbreak that probably gained them a grand total of one extra acre of cropland. Had these same farmers gone to a land auction where someone paid $20,000 an acre for land, they would be guffawing about it at the coffee shop for decades to come.

The bigger tragedy is that this expensive decision actually cost them yield. Research at the University of Nebraska experiment field at Mead has shown that crops protected by windbreaks around every 40 acres out-yield those in the open by 20 percent for corn, 18 percent for soybeans, and 22 percent for wheat. It has been determined that, in eastern Nebraska, windbreaks could take up as much as 5 percent of the acreage and be more profitable than the same land in use without windbreaks.

You can also reduce wind speed at the soil surface by leaving tall stubble heights of crop residue in fields or by leaving tall stubble heights after cattle have grazed in pastures. The data on the facing page show how standing stubble and narrow strips of tall wheatgrass arranged perpendicular to the prevailing wind improve soil moisture in the Great Plains region.

Much of the value of wind speed reduction in the Plains is thanks to the trapping of windblown snow. Trapped snow can be a significant addition to soil moisture, and the measures that improve snow capture in the winter are the same ones that reduce wind speed in the summer.

How Temporary Windbreaks of Standing Corn Affect Soybean Yield

Showing the average of 11 paired site-years in Minnesota, this data illustrates how strongly wind-breaks of mere standing corn can influence crop yields.

YIELD IN OPEN AREA	YIELD IN AREA PROTECTED BY WINDBREAK	INCREASE
24.7	27.4	11%

Table 3.3

How Wind Barrier Strips Affect Soil Moisture

From a study tracking the effectiveness of wheat stubble and tall wheatgrass wind barrier strips in Montana, the following is the average of results during 8 years.

	MOISTURE WITHOUT BARRIER STRIPS	MOISTURE WITH BARRIER STRIPS
Bare fallow soil	2.8" (7.1 cm)	4.1" (10.4 cm)
Standing stubble	4.5" (11.4 cm)	6.5" (16.5 cm)

Table 3.4

Windbreaks or shelterbreaks of trees decrease wind speed at the soil surface and reduce evaporation of soil moisture.

Protecting Soil from Excessive Sun and Heat

Sunlight striking bare soil and high soil temperature are both prevented by keeping the soil covered with either live plants or dead ones. Live plants do use moisture through transpiration, but if those plants are desirable, this use is necessary and unavoidable. Dead plants are the residue left over from previous crops. If your goal is to conserve moisture, you should never mechanically harvest crop residue.

Leaving taller stubble heights after harvest is also beneficial. One way to help the crop residue preserve soil moisture is to harvest cereal grains with a stripper header rather than with a platform header. According to data from the University of Nebraska (see References), wheat stubble harvested with a stripper header averaged 9 inches of available moisture on April 12 of the spring following the wheat harvest, while platform-harvested wheat stubble contained 4.3 inches on the same date, due to taller stubble height left by the stripper header.

If there is insufficient plant cover in your cropland, establish cover crops. You might object to this, especially if a drought is already underway. A cover crop does use moisture, despite what some overzealous cover crop advocates may claim. Once the cover crop is dead, however, and rainfall replenishes the moisture used by the cover crop, that field is much more moisture-efficient going forward.

Cover cropping can be a tremendous tool to improve moisture efficiency of cropland, but if you wait until the drought is underway to begin cover cropping, you will realize much less benefit compared to those who begin cover cropping in favorable years and benefit from the already present mulch in later dry years. Trying to grow a mulch when the soil is bone-dry and has no cover is usually

Corn mulched with hairy vetch (on the left) compared to corn grown without mulch (on the right)

How Tillage Method Affects Rainwater Storage

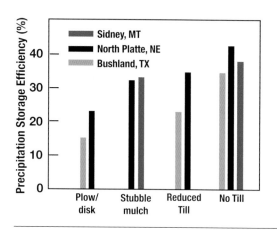

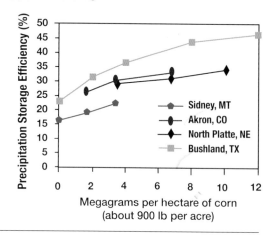

Table 3.5

Research at the USDA Agriculture Research Center at Akron, Colorado, shows that as the tillage intensity during a fallow period increased, the amount of stored moisture decreased.

disappointing, just like trying to grow a cash crop under the same conditions.

In pastureland, you may be able to inter-seed short-lived annuals into overgrazed pasture to provide off-season cover — for example, planting winter clovers into dormant native grass or bermudagrass in the fall. This can take advantage of off-season rains to produce additional soil cover and/or feed. Be sure to select species that will not become a weed that competes with the perennial pasture species. This concept is discussed further in chapter 6.

IMPROVING THE WATER-HOLDING CAPACITY OF THE SOIL PROFILE

The amount of water a soil can provide to plants is determined by the water-holding capacity of the soil per unit of depth as well as the depth to which roots can penetrate to extract the water. Methods for increasing root depth are explored in chapter 4.

Leaching is the movement of water below the root zone, as occurs when more water moves into the soil than the soil can hold. The water-holding capacity of a soil is determined by two primary factors:

Soil texture. This describes the proportions of sand, silt, and clay in the soil. Sand has very little ability to retain water, while clay has excellent water-holding capacity (at least the clay materials found in temperate regions). Silt is intermediate.

Organic matter in the soil. Organic matter can dramatically improve soil water-holding capacity, and pound for pound it has a far greater impact than clay.

Table 3.6 on the next page illustrates how both texture and organic matter affect soil water-holding capacity. Obviously, it is prohibitively expensive to alter the percentage of sand, silt, and clay in the soil on a large scale. It is possible, however, to improve the amount of organic matter in a soil; therefore, we will focus on organic matter.

How Organic Matter Affects Soil's Water-Holding Capacity

Increasing the soil organic matter from 1 percent to 5 percent on a silt loam can more than double its water-holding capacity. Data is in inches of water per foot of soil.

% ORGANIC MATTER	SAND	SILT LOAM	SILTY CLAY LOAM
1	1.0	1.9	1.4
2	1.4	2.4	1.8
3	1.7	2.9	2.2
4	2.1	3.5	2.6
5	2.5	4.0	3.0

Table 3.6

An additional perk is that as organic matter increases so does soil aggregation. This improves oxygen penetration into the soil, which in turn improves root depth. For example, a silt loam soil with organic matter of 1 percent (holding 1.9 inches per foot) can hold 3.8 inches of available water when saturated. The same soil with 5 percent organic matter (holding 4 inches per foot of depth) can hold a total of 16 inches of available water, once saturated.

To illustrate how much difference that can make, let's assume that the ground is planted to corn. Since a crop of corn will use about 0.3 inches of water per day from tasseling to maturity, the low-organic, poor rooting-depth soil will hold only about 13 days' worth of moisture until wilting occurs, while the high-organic-matter soil with the better rooting depth can hold 53 days of water before wilting. Since corn usually takes about 50 to 55 days from tassel to maturity, if the soil is full at tasseling the crop can make it to maturity without stress even if no rain falls.

PUTTING A DOLLAR VALUE ON WATER-HOLDING CAPACITY

So what is the additional water held by that organic matter worth? Obviously, it depends on the degree to which water is the limiting factor of production. In my area in Kansas, lack of water occurs virtually every growing season at some time or another. If you live where it rains every day, additional water-holding capacity has little value. In areas that receive infrequent but large rains, like much of the Great Plains, the ability to store this moisture may be priceless.

How do you put a value on water-holding capacity? One way is to determine how much each additional inch of water is worth in terms of crop yield. Each crop requires a certain amount of moisture to get to the first unit of yield, called the **threshold water use**. Corn, for example, requires a certain amount of moisture to grow the plant to the reproductive stage. After that water is invested, each additional inch of water goes into grain production. Since with forage plants we harvest the whole plant (or at least a large percentage of it), even small amounts of water can produce a harvestable yield.

The threshold water use plus the yield per each additional inch are listed on the facing page for several crops. Keep in mind that it takes more water to produce a bushel of corn in western Texas than it does in Minnesota. Higher heat, lower humidity, and higher wind speed all increase these numbers. Many of these values were determined by the USDA Agricultural Research Service (ARS) in Akron,

How Additional Rainfall Can Affect Yield

CROP	THRESHOLD WATER USE	ADDED YIELD PER INCH OF MOISTURE
Corn	10.9 inches	16.9 bushels
Grain sorghum	6.9 inches	12.2 bushels
Sunflower	5.4 inches	218 pounds
Winter wheat	10.0 inches	6.0 bushels
Soybeans	7.8 inches	4.6 bushels

Table 3.7

Colorado, which is located in the eastern part of the state in semi-arid shortgrass plains. You should adjust the given numbers according to your climate.

With the information in Table 3.7 it becomes feasible to assign an approximate economic value to added organic matter. Suppose you are a corn grower. If an additional percentage point of organic matter holds 0.5 inch of added moisture, and each additional inch of moisture adds 16.9 bushels of corn every year that moisture is limiting (see box on the next page), then that percentage point increase in organic matter is worth 8.45 bushels of corn times the price of a bushel, each and every year in which water is a

The Law of the Most Limiting Factor

Justis von Liebig proposed a theory in which plant growth proceeds so far as all the essential growth factors are present, until one of the growth factors is no longer sufficient. At this point, growth ceases, no matter how well supplied the plant may be with all the other growth factors. This first growth factor that becomes inadequate in supply becomes "the most limiting factor" and should be the focal point of management.

For example, if there is enough sunlight to produce 20 tons of forage per acre, enough water to produce 10 tons of forage per acre, enough nitrogen to produce 3 tons per acre, and all other growth factors are sufficient to produce 30 tons per acre, the yield will be no more than 3 tons; the yield is limited to the amount of available nitrogen. It is the most limiting factor. In this situation, addition of nitrogen to the system should be the goal of the land manager, if higher yield is desired.

If enough nitrogen is added to produce 30 tons, the next most limiting factor will become water, and the yield will now become 10 tons. Now the focus of the manager is to improve soil-water relations. It is important to note, however, that the most limiting factor can change during the season, perhaps even hourly. In early spring, there may be plenty of moisture, but perhaps plant growth is limited by lack of phosphorus, or by lack of heat. A month later, nitrogen may be the limiting factor, and a month after that it may be water. Application of growth factors (fertilizer, irrigation water) will result in yield increases only if that growth factor is limiting at some point in the season.

limiting factor, in water-holding capacity alone. So if corn is worth $5 a bushel, an additional percent organic matter that gives 8.45 bushels worth of water stress alleviation is worth $42.25 a year per acre.

Of course, organic matter offers more value than just water-holding capacity, so don't get the impression that this is the total of the economic value of organic matter. We've been looking at the water-holding capacity alone, but added organic matter also almost always improves root depth. As might be expected, added root depth is also extremely effective at increasing water-holding capacity.

HOW TO INCREASE SOIL ORGANIC MATTER

In college I was told that while it is possible to increase soil organic matter, it would never be economically feasible to do so. That assumption arose from research indicating that even huge additions of manure could not permanently increase organic matter content on *tilled* soil. At the time, very little no-till farming was practiced. An article in *Agribook* magazine in 1977 described the apparent futility of trying to increase soil organic matter. It claimed that by adding livestock manure at the rate of 10 tons per acre per year, it would take 75 years to increase soil organic matter by one percentage point.

The psychological effect of that statement, and many others like it from the same era, was to convince generations of farmers that increasing soil organic matter is an unachievable goal, so don't even bother to try. However, no one really investigated what was possible with the combination of manure applications and no-till. Likewise, cover cropping was not a widespread practice at that time, so no one investigated the combination of no-till, manure, and cover crops. And while mycorrhizal fungi were known to most soil scientists, commercial inoculants were not available, so no one looked at any combination of no-till and mycorrhizal fungi. Obviously, no one investigated a combination of all four methods: no-till, cover crops, mycorrhizal inoculation, and manure. I have combined all these on the same field and found the results can be absolutely dramatic.

Due to the dynamics of soil organic matter, every year a certain amount of organic matter breaks down naturally. A soil deprived of organic matter input gradually declines in organic matter over time. Ordinarily, around 2 percent of the organic matter in a soil will break down each year. This amount is higher in a sandy soil than in a clay soil, and it's higher in a hot soil than a cold soil. If a soil is 4 percent organic matter and the surface 6 inches weighs 2 million pounds, there are 80,000 pounds of organic matter per acre. If 2 percent of that breaks down each year, that means we need to replace 1,600 pounds of humus each year just to break even. If we fail to do that, our organic matter decreases, and if we exceed that amount the organic matter increases.

Since we want increases, let's look at specific ways to improve soil organic matter.

1. ELIMINATE TILLAGE

While it is still a matter of argument about whether or not tillage increases crop yields, there is absolutely no question about the effect of tillage on soil organic matter: it decreases it (see Table 3.8). The evidence is crystal-clear. While no-till alone usually only slowly increases soil organic matter, it is a necessary first step, without which other measures on cropland are much less effective. Even though the organic matter increases are usually small, the increase in water-holding capacity can be large over time, as illustrated in Table 3.9.

2. MAXIMIZE CROP YIELD

Organic matter comes from plant biomass. Therefore, it is reasonable to expect that better-yielding crops (whether they are grain crops, vegetables, or pasture) will contribute more to soil organic matter than lesser-yielding crops. Removal of barriers to yield, such as poor soil fertility, helps increase soil organic matter (see Table 3.10). While it seems like this would be something everyone would do anyway, in many situations it is not done, particularly on pastures.

3. RETAIN CROP RESIDUES

When hay prices are high it is often tempting to bale up crop residues and sell them, or to use them as feed and sell hay. In chapter 2, I pointed out the folly of baling crop residue. Leaving crop residue on the field also pays long-term dividends in better organic matter. Table 3.11 illustrates the effect of crop residue management on soil organic matter.

While it is hardly surprising that removing crop residue for several consecutive years will reduce soil organic matter, the degree to which it is affected should be rather shocking.

Grazing crop residue rather than baling still allows the use of this material as feed with the added advantage of manure return. If crop residues are baled, they should be fed in the field to ensure manure return.

4. ADD COVER CROPS

If the residue of one crop a year is good, then the residue of two crops a year is even better. Cover crops usually have an even more pronounced effect on organic matter than the residue of the cash crop because they are typically killed in a vegetative state while the foliage is still high in protein. Thus, they have a more favorable carbon-nitrogen ratio. This means they form more humus per pound of residue than crop residue does, and they can provide some protein to microbes at work degrading adjacent crop residue. Adding a cover crop to the total organic matter pool is like adding the income from a second job to the household cash flow.

How Long-Term Tillage Affects Soil Organic Matter

Research from Missouri indicates that as tillage becomes more intense, soil organic matter decreases, as shown below.

TILLAGE SYSTEM	PERCENT SOIL ORGANIC MATTER
Moldboard plow	2.5
Chisel	3.3
Disk	3.5
No-till	4.1

Table 3.8

Typically, the higher the biomass generated by the cover crop, the more favorable the effect upon soil organic matter. Therefore, high-biomass crops like sudangrass have more effect than low-residue crops like buckwheat. The graph below illustrates the effect of cover crops on soil organic matter.

How Long-Term Tillage Affects Plant-Available Water Capacity

These data are conclusions from long-term experiments in New York state.

SOIL	DURATION	MOLDBOARD PLOW	NO-TILL	% INCREASE DUE TO NO-TILL
Silt loam	33 years	24.4	28.5	17%
Silt loam	13 years	14.9	19.9	34%
Clay loam	13 years	16.0	20.2	26%

Table 3.9

How Nitrogen Fertilization Affects Crop Yield and Organic Matter

The data below illustrate how strategies to achieve high yields can also add to soil organic matter.

N APPLIED (LB/ACRE)	CORN GRAIN YIELD, 12-YEAR AVERAGE (BU/ACRE)	SOIL ORGANIC MATTER AT END OF 12 YEARS
0	39.1	2.70
60	82.1	2.85
179	131.5	2.90

Table 3.10

How Residue Management Affects Soil Organic Matter

Leaving residue on the field will increase organic matter, as shown in this data, tracking a continuous crop of corn in Indiana.

TREATMENT	% SOIL ORGANIC MATTER AFTER 6 YEARS
Fallowed (no crop grown)	2.60
All crop residue removed	2.75
Residues returned	3.05
Double residue returned	3.36

Table 3.11

How Crop Residue Affects Soil Moisture

As the amount of crop residue increases, so does the level of soil moisture.

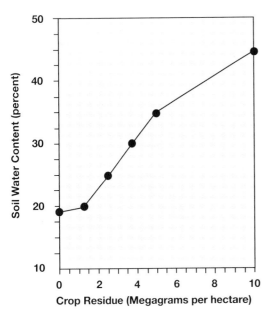

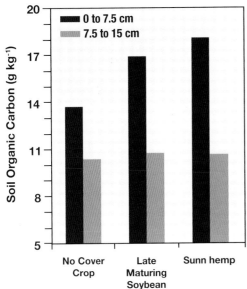

Table 3.12

5. OBSERVE THE CARBON-NITROGEN RATIO

It takes more than carbon to make humus: it also requires nitrogen, as well as sulfur and several other minerals, such as phosphorus. If soil is properly fertilized to meet crop need, there should never be a deficiency of *most* of these nutrients that would limit humus formation. Nevertheless, it is quite common for crop residue to have insufficient nitrogen and sulfur for humus formation.

Humus has a ratio of about 1 part sulfur to 10 parts nitrogen to 100 parts carbon. If any of these components is lacking, the amount of humus formed is less than optimal. Since the nitrogen and sulfur in plant residue reside largely in the protein, the protein content of a residue is the best clue as to whether it has enough nitrogen and sulfur for optimal humus production. If a residue has at least 12 percent protein, it will produce the maximum amount of humus from the residue. If it has much more than 12 percent protein, it will also release available nitrogen, acting as a green manure. If a residue has less than about 12 percent protein, its decay will not only form very little humus, but it will also tie up soil nitrogen. If this is the case, add sulfur and nitrogen to improve humus formation, preferably in an organic form such as manure that also provides biological energy and a humectant effect.

◁ Kansas State University research shows that soil organic matter increased when cover crops of sunn hemp or forage soybean were added to a sorghum-wheat rotation. The cover crop was planted in the wheat stubble in a two-year rotation, totaling three treatments in the study's six-year duration.

The most productive crops in terms of biomass per acre are warm-season grasses like corn and sorghum. Their residue is low in protein, however, and requires either blending with high-protein residue or addition of fertilizer to balance the carbon-nitrogen ratio. The problem is that a microbe needs both carbon and protein at the same place at the same time. It is difficult in practice in a field to have the nitrogen and residue mixed intimately enough so that a microbe (able to travel only a few microns in its very short lifespan) can access both.

Ideally, the residue and nitrogen source would be finely ground, mixed together, and kept warm and moist for the microbes to do their magic. This happens only occasionally in the soil, but it happens 24 hours a day, 365 days a year in the rumen of a cow or other ruminant. That's why there is such magic in animal manure, particularly in the manure of ruminants.

One experiment station in Kansas conducted a long-term study in which one plot had additional crop residues added to it to equal twice what it produced. At the end of this study, it yielded no better than the control plot with no added residue. Another study, unrelated to the first, measured the effect of adding a one-time application of cattle manure (ironically, the amount of manure was close to what you would get by feeding the crop residue from the previous study to the cattle), and this increased crop yields by a tremendous amount (about 25 percent) over the duration of the study. Needless to say, converting plant material into manure is a desirable process. A third study, conducted in Guelph, Ontario, (below) had similar results.

As shown in this data, the return of manure had a similar effect on soil organic matter as the return of the residue, but it had a much more positive impact on crop yield. One explanation for the difference is that the residue can act as a host for diseases and has an unfavorable carbon-nitrogen ratio, so it can tie up nitrogen, thus often temporarily depressing yields. The fermentation action inside the rumen eliminates both of these issues, to the benefit of the following crop.

How Organic Matter with Residue Treatments Affects Crop Yield

In this Ontario study, residue was either removed, left on the soil ("returned"), or removed and fed to livestock, with the manure produced by the livestock returned to the field.

RESIDUE	REMOVED	RETURNED	REMOVED BUT MANURE EQUIVALENT RETURNED
% organic matter	3.7	3.9	3.9
Yield, bu/acre	80.4	78.6	91.4

Table 3.13

6. APPLY MANURE

One of the most effective ways to improve soil organic matter is to apply livestock manure. The most common means to do so is to obtain it from a confined livestock operation and mechanically spread it. A better option is spaced bale feeding or bale grazing as discussed in chapter 2. By whatever means the manure gets in the field, its benefit to the soil is quite dramatic, as shown in Table 3.14.

7. PLANT PERENNIAL FORAGES AND GRAZE THEM

The absolute best crop to plant for improving soil organic matter is a blend of perennial grasses and legumes, particularly warm-season grasses like big bluestem. A blend of grasses and legumes together is superior to grass alone. Grazing is far superior to haying as a harvest method, unless the hay is fed right on the field where it was produced to return the manure back to the field. This additional organic matter can significantly improve the soil water-holding capacity, as illustrated in Table 3.15.

8. INOCULATE WITH MYCORRHIZAL FUNGI

The **hyphae** (filaments) of mycorrhizal fungi secrete **glomalin**, a form of stable organic matter. Mycorrhizal-colonized plants are far more productive of plant biomass, root exudates, and soil organic matter than plants that are not colonized, particularly on perennial forages. But the biggest impacts of mycorrhizal fungi result from the ability of glomalin to hold water itself, and its ability to produce water-stable soil aggregates (see Table 3.16), which increases root depth, as discussed in chapter 4.

How Manure Application Affects Soil Organic Matter

This table shows the effect of 11 years of manure additions on soil properties of a clay soil in Vermont with continuous corn silage and conventional tillage. The original soil organic matter at the beginning of the experiment was 5.2 percent, so it took 20 tons of manure per acre annually in this environment to maintain soil organic matter.

EFFECT ON SOIL	BEGINNING ORGANIC MATTER	NONE APPLIED	10 TONS/ ACRE	20 TONS/ ACRE	30 TONS/ ACRE
Final soil organic matter level	5.2%	4.3%	4.8%	5.2%	5.5%
Total pore space		44.0%	45.0%	47.0%	50.0%

Table 3.14

How a Cropping System Affects Water Capacity

SOIL TYPE, PLANT	AVAILABLE WATER CAPACITY AFTER 12 YEARS
Loamy sand, continuous corn	14.5%
Loamy sand, corn after grass	15.4% (6% increase)
Sandy clay, continuous corn	17.5%
Sandy clay, corn after grass	21.3% (22% increase)

Table 3.15

How Glomalin Affects Water-Stable Aggregates

In this study, glomalin levels increased under no-till compared to tilled systems, and with perennial crops compared to annuals. As glomalin increased, so did the percentage of water-stable aggregates.

CROPPING SYSTEM	TOTAL GLOMALIN (UNIT?)	% OF AGGREGATES THAT ARE WATER-STABLE
Spring wheat to fallow, conventional tillage	2.7	23
Spring wheat to winter wheat to safflower, no-till	3.8	58
Moderately grazed pasture	7.9	93

Table 3.16

9. ADD BIOCHAR

Biochar is a particularly effective additive for improving water-holding capacity. It is charcoal that has been infused with a liquid containing mineral nutrients and beneficial microbes, such as manure. The charcoal has a huge number of microscopic pores, the remnants of old cell walls. These pores and the internal surface area associated with them offer a great deal of water-holding capacity per pound of biochar.

How to Apply Biochar

It takes a lot of biochar (several hundred pounds per acre) to have a significant impact on total water-holding capacity, and adding this quantity usually is not economically feasible on most crops. Adding a little at a time does add up, however, and if it's applied in a manner that gives the best "bang for the buck" it can pay for itself. For example, higher-value application methods include as a seed coating, or as an in-furrow additive, where biochar acts to enhance germination.

Another method (my favorite) is to use biochar as a feed supplement, so that it gets infused with nutrients and microbes in the digestive tract and is deposited in the manure. This has multiple benefits, in addition to long-term soil improvement. If the animal is a ruminant, the natural formation of methane in the rumen results in a substantial loss of feed energy when the animal belches out the methane. There is evidence that biochar can **adsorb** (strongly attach to) methane and keep it from being lost via belching. Rumen microbes then use the methane for energy, and the animal then digests these microbes, thus increasing feed efficiency by turning what was formerly wasted (methane) into a feedstuff.

After the manure is deposited, the biochar adsorbs ammonia, keeping it from volatilizing into the air. Since up to half of the ammonia in a cowpie can be lost to volatilization in as little as 72 hours after deposition, this can significantly add to the fertility value of the manure. A pound of biochar can trap up to a half pound of ammonia. Furthermore, biochar

How Biochar Affects Soil Water-Holding Capacity

TREATMENT	AVAILABLE WATER %	INCREASE OVER CONTROL
Control	0.242	NA
Deep banded biochar at 3 percent of soil by weight	0.274	13%
Biochar mixed into soil at 3 percent by weight	0.285	18%
Deep banded biochar at 6 percent of soil by weight	0.276	14%
Biochar mixed into soil at 6 percent by weight	0.3	25%

Table 3.17

can adsorb toxins, including botulism toxins (found when dead animals contaminate hay), toxins found in moldy hay, and toxins produced by intestinal pathogens, such as calf scours caused by *Cryptosporidium parvum*. These benefits essentially pay for the biochar, and the long-term soil improvement aspects can be considered a nice bonus.

One other benefit to biochar is odor reduction in manure, especially in hogs. If you have lawsuit-happy neighbors, that may turn out to be a big deal. In any event, having a better-smelling farm can make it a more pleasant place to live.

Adding biochar at 4 ounces per day per cow in the minerals can pay for itself in improved feed efficiency and retained nitrogen fertility. At a **stocking rate** of 1 cow per acre for 150 days, this would add about 38 pounds of biochar per acre. At this rate, it takes years to get the per-acre amount needed to make a

big difference in water-holding capacity, but the added biochar is paying its way as it goes.

If this technique is combined with bale grazing or grain feeding on pasture, in which a large number of animals can be kept on a small acreage because feed is being imported to the area, the amount of biochar that can be deposited can be dramatically increased. For example, if bale grazing is used on a field, 10 cows per acre can be kept for a month, resulting in a deposit of 76 pounds of biochar. If using spaced bale feeding, in which bales are placed on a grid 30 feet apart in all directions, 50 cows can be kept on an acre for a month, resulting in a deposit of 380 pounds of biochar per acre. Now we are getting up there to levels that make a difference.

Another method that deposits a large amount of manure in a short period of time is to feed biochar in with the grain supplement of pastured hogs or poultry.

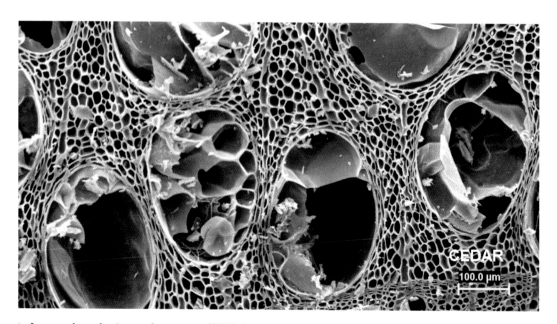

A scanning electron microscope (SEM) image of the cellular structure of biochar (in this case, cedar)

Adding biochar via manure can have other significant effects on soil, particularly on the availability of mineral nutrients. In an Australian study (see References), biochar was fed to pastured cattle for 3 years at a rate of 0.33 kg (about 12 ounces) per head per day, mixed with molasses. The carbon content of the soil went from 41.7 to 46.7, the soil-test phosphorus went from 49 to 202, and the soil-test potassium went from 55 to 205. This is a dramatic improvement in nutrient availability.

Biochar Caveats

Note that biochar adds few, if any, minerals. The improvement in nutrient availability demonstrated in the Australian study is an increase in availability of what the soil already contains, presumably due to the favorable effect biochar has on soil microbial activity when in intimate contact with manure.

One drawback of the "feed through" method of putting biochar in livestock feed is that it is not an approved feedstuff in the United States, despite being used in animal feed for almost two centuries by farmers. It is currently illegal for feed manufacturers to put it into manufactured feed. Apparently, there are no prohibitions from farmers doing so, as it is a naturally produced product.

Obviously, biochar can add tremendously to the ability of the soil to hold water (see Table 3.17 on page 62). However, note the amount used to make these changes in the study cited. If the top 6 inches of an acre weighs 2 million pounds (a standard assumption), then 3 percent of that amount by weight is 30 tons. Applying that amount of purchased biochar is completely uneconomical on a field crop, though it may be used on a garden scale.

My intention here is not to discourage the use of biochar as a soil amendment; far from it. Biochar can be very valuable, even at far lesser amounts than those in the research above, and I am a biochar fan and strongly encourage its use. I just want to avoid unrealistic expectations of complete soil transformations with the application of affordable amounts per acre of biochar, because many of the eye-popping examples of tremendous gains in soil quality by biochar application involve very large amounts. It is a legitimate soil amendment, and if used in a manner that pays its way as it goes (as a seed treatment, in-furrow application, feed ingredient, and so on), the effects are cumulative over time and can eventually be quite beneficial. There are numerous examples where far lesser amounts of biochar than what was applied in the study above have had very beneficial results in crop yield and soil quality.

Heat: Biochar's Other Benefit

Biochar can be manufactured so that the heat produced is utilized for benefit, such as with cooking stoves that produce biochar, or biochar kilns used to heat buildings or greenhouses. This means that biochar is not only a great benefit to soil, it also can replace the use of fossil fuels for an additional benefit to the planet.

CHAPTER SUMMARY

There are multiple ways to improve the ability of your soil to retain available water. In combination, they can provide enough effect to get most plants through most drought events.

- Transpiration is a necessary water loss if you are growing plants, but transpiration by weeds should be eliminated. The best way to eliminate a weed is to convert it into a beneficial plant, such as by finding a livestock species that will eat it or by training livestock to eat it.

- Evaporation can be reduced by reducing wind speed and by keeping soil covered.

- Wind speed can be reduced by leaving stubble standing and by installing windbreaks.

- Soil cover can be increased by leaving crop residue on the field rather than removing it or tilling it under.

- Cover crops can add soil cover and additional organic matter, and they can improve the carbon-nitrogen ratio of the residue layer.

- The ratio of sulfur to nitrogen to carbon in the residue layer determines how much humus can be produced from it. Most of the sulfur and nitrogen in a plant is in the form of protein; for optimum humus production a residue should be around 12 percent protein.

- Manure can dramatically improve soil water-holding capacity. You can generate manure by letting animals graze the vegetation produced on an area. To obtain a greater amount you can bring it in from external sources such as confinement livestock operations and spread it mechanically, or it can be self-spread by animals fed imported feed.

- The fastest route to increased soil organic matter without the importation of organic material is to plant perennial pasture blends of grasses, legumes, and forbs, and graze them.

- Mycorrhizal fungi produce a strong soil-aggregating agent called glomalin that greatly improves soil water-holding capacity. Inoculate crops with mycorrhizal fungi and then keep the fungus alive by maintaining constant live roots.

- Biochar can add greatly to soil water-holding capacity. The most profitable use of biochar may be as a livestock feed supplement.

4

HELPING PLANTS GET MORE WATER OUT OF THE SOIL

Whenever the subject of improving root depth comes up, the first thing that tends to pop into a farmer's mind is getting out their biggest tractor, hooking it onto a subsoiler, and tearing the soil up just as deep as they possibly can.

Surprisingly, subsoiling is largely ineffective at improving rooting depth. I raise more than a few eyebrows when I make that statement, but most of the research trials I have seen back me up.

THE SUBSOILING MYTH

The conversation goes like this: Someone will tell me, "But I subsoil and I always get a yield response of a bazillion bushels, and my soil is so loose and mellow [etc., etc.]." Then I always ask, "Did you leave a control strip where you did not subsoil, and weigh the yield from each area separately?" The answer is almost always "Uh, no." Then I ask, "Did you dig a pit in the field to see if your roots actually went deeper?" Again, the answer is always no. But farmer after farmer tells me subsoiling improves their yields, when university research in the same geographical area says the opposite.

Why the discrepancy? I believe I have hit on the answer. I found a research trial that tracked CO_2 evolution (a measure of the amount of organic matter breakdown) after various tillage trips, and the CO_2 evolution from a subsoil trip was enormous, similar to that from moldboard plowing. If organic matter is 5 percent nitrogen, the breakdown of this huge amount of organic matter results in a large amount of nitrogen being converted from stored reserves into available nitrogen. It acts just like applying extra fertilizer.

Despite the popular media portrayal of farmers dumping excess fertilizer on their fields, I see that most of them routinely under-fertilize, trying to save a few dollars per acre. By contrast, experiment fields are almost always adequately fertilized. Because fertility is often a limiting factor in farm fields, the extra nitrogen released by the subsoil trip increases yield there, but it does not increase yield in the adequately fertilized experiment fields.

Studies repeatedly show that subsoiling is not a productive use of time or resources.

The Research Doesn't Back It Up

The research trials I have seen that have actually measured root growth (see References) almost always indicate that the more intense and deeper the tillage event, the *less* root growth there is. This is because tillage does not permanently improve (and actually worsens) the real factor that limits root depth.

- In a University of Kentucky Extension study, corn after subsoiling yielded 161.6 bushels per acre, while corn without subsoiling yielded 160.3 — hardly an economical response to such an expensive operation.

- In an extensive study conducted in Iowa and Illinois, subsoiling failed to improve yields over multiple sites and multiple years.

- Research at the USDA-ARS in Waseca, Minnesota, showed not only no improvement due to subsoiling but also an 11-bushel decrease in corn yield over the year the trial was conducted.

The list of trials showing no value to subsoiling goes on and on.

One exception to this occurs in the southeastern United States, where there does seem to be a significant yield increase from subsoiling. This may be due to the lack of frost action during the winter in these areas. Even in this environment, however, the yield increase lasts only one year, and the subsoiling must be repeated annually to provide benefit.

There is a need for long-term research in the Southeast comparing subsoiling to other methods of alleviating compaction, like the ones discussed in this chapter. I contend that other methods will eventually produce much better long-term effects at much lower costs.

THE REAL LIMITING FACTOR TO ROOT DEPTH

When you ask people what keeps their roots from going deeper, the answer is almost always, "Well, it's because my soil is so %$#$#%$ hard." Surprisingly, that is not the reason. Have you ever walked down a city sidewalk and seen where tree roots have broken the concrete? Plant roots can accomplish amazing things, even break concrete, if they have all the growth factors they need.

I have a friend who once hauled sand in his grain truck just after hauling soybeans to the elevator. He put down the sand as a base prior to pouring a concrete floor for a grain bin. Before the concrete had fully cured, he noticed soybean seedlings popping divots of concrete out of his floor. Even soybean seedlings can break concrete. So how can a thin layer of plowpan be so hard that it stops roots, when concrete cannot? The truth is that a buried layer of plastic film would stop roots as effectively as that hardpan, and for the same reason.

The true reason roots can't penetrate either plastic film or a plowpan is because both are barriers to the real limiting factor to root depth, and that is oxygen penetration. Roots require a minimum of 10 percent oxygen in order to function. Atmospheric air is 21 percent oxygen, and as it penetrates the soil, the oxygen gets used up by soil microbes and plant roots, and the deeper you go the less oxygen there is. Obviously, the depth where oxygen content drops below 10 percent is where the root growth stops.

The challenge, therefore, is how to increase oxygen penetration into the soil.

Subsoiling can reduce mechanical resistance, but mechanical resistance isn't the problem, and since it breaks down organic matter and pulverizes aggregates, it actually reduces the penetration of oxygen into the soil over the long term.

It is easy to determine the average oxygen status of soil by simply looking at the color of the subsoil. Most subsoils are rich in iron compounds. If these compounds are exposed to oxygen on a routine basis, they oxidize to red-colored compounds, chemically similar to rust. If there is a lack of oxygen, the subsoil will be gray. Intermittent or incomplete oxidation results in either a yellow subsoil or a mottled mixture of reds and grays.

Subsoil that is well oxygenated will have a red color (left), while a subsoil with poor oxygen content will have a gray color (right). The well-aerated soil was taken from a native grass (never plowed) pasture, while the poorly aerated soil was taken across the fenceline in historically tilled cropland. The core pieces shown were taken from a depth of three feet.

INVITING OXYGEN INTO THE SOIL

So how can we improve penetration of oxygen into the soil? Most of the methods are similar to those used to increase water penetration into the soil: earthworm channels, root channels, and improved aggregation (these methods are addressed in chapters 2 and 3). The problem is, while earthworm burrows and old root channels can conduct more oxygen into the soil, earthworms and roots need oxygen to grow in the first place. So how do you get the very first root to penetrate the oxygen-limiting layer? There are a couple little tricks to accomplish this.

MANIPULATING WATER AND TEMPERATURE

Oxygen levels fluctuate in the soil with temperature and moisture levels. The more water in the soil, the less oxygen there is. The pore space is limited in the soil, and air and water cannot occupy the same space at the same time. However, managing moisture levels to manipulate oxygen levels is pretty difficult. Adding water is easy if you have irrigation, but taking water out of the soil to improve oxygen is tough.

A solution is subsurface tile drainage. Its goal is not to remove excess water but rather to introduce oxygen into the root zone. Contrary to what you might think, subsurface drainage can actually create a more drought-tolerant field — allowing full root growth by eliminating oxygen deficiency in the subsoil. Installation of tile drainage (which is a bit of a misnomer now, as most drainage is now accomplished with perforated plastic tubing rather than the clay tiles formerly used) has made thousands of acres across the

Midwest vastly more productive by allowing full root growth on formerly low-oxygen soils.

Temperature, on the other hand, is hard to manipulate but is quite easy to predict. It goes down in the winter and up in the summer. And as temperature goes up, oxygen level goes down. This is because the soil microbiological activity goes up with the temperature, and soil microbes use oxygen.

Soil microbes become inactive when soil temperature drops below 50°F (10°C). Therefore, the easiest time for roots to penetrate a plowpan is when soil temperature is below that point but above freezing, because there is no microbial competition for oxygen. In temperate climates, there is a short time period in the spring when this occurs, and

How Cover Cropping Affects Corn Root Growth

In this study, corn grown after rye or forage radish had a greater number of roots at every sampling depth in the soil, with radish being particularly effective at improving root density below a depth of 12 inches.

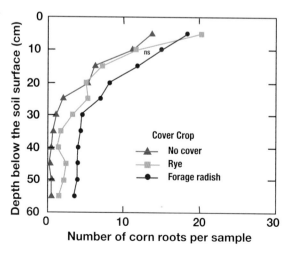

Table 4.1

a somewhat longer period in the fall. Plants that can grow a large volume of deep roots in the fall are thus ideal for this task. Summer annual crops, like corn and soybeans, have no active roots during these time periods, which is a major reason behind the tendency of soil to become increasingly compacted in a corn-soybean rotation.

Some of the very best crops for breaking up a root-limiting layer include the brassicas, like canola and forage radish, as they have deep taproots that grow very well in fall. The recent increase in popularity of cover cropping throughout much of the United States can be largely attributed to the promotion of Tillage radish, a brand name for a variety of forage radish. It is popular because it works. When planted approximately 60 days before the average first frost, forage radishes can develop extremely large taproots that can penetrate up to 2 feet (0.61 m) deep. Left in place to winterkill, which they do when temperatures drop below 16°F (−8.8°C) for several consecutive nights, they will rot with warmer spring temperatures and leave a root channel the size of a gopher hole (see photographs on page 28).

After this process, it is critical that the next crop be planted no-till. Even a little tillage will disrupt the connection between the surface and the subsoil for oxygen to flow through. Remember, there is a big difference

Compaction-Busting Plants

Several radishes on the market are just as good as the Tillage radish, notably the Nitro radish. Other brassicas that are similarly effective, though to a lesser extent, include canola (which can be harvested as a cash crop), hybrid brassicas (such as Winfred), forage collards, turnips, and kale.

Keith Berns poses with 18-month-old daughter and 3-month-old radish. (Photo courtesy Green Cover Seed)

Yellow blossom sweetclover (*Melilotus officinalis*) was once the most popular "green manure crop" in the Corn Belt, valued for both its ability to fix nitrogen and its deep taproots that could break through plowpans.

between having your snorkel an inch above versus an inch below the water surface.

THE SWEETHEART OF COVER CROPS

Sweetclover is a legume with very deep taproots. It is usually spring-planted, then left to grow for a calendar year, being killed prior to the planting of another crop the following spring.

Sweetclover was once the most common cover crop (or green manure crop) grown in the United States. For decades, the standard crop rotation in the Corn Belt was corn, followed by oats with an underseeding of sweetclover. The sweetclover was allowed to grow after the oat harvest until the following spring. In other areas, red clover was the preferred cover crop. Red clover produces forage of a much higher quality than sweetclover, but of lower tonnage.

Sweetclover is able to grow on poorer soils (except acidic soils), is a better nitrogen fixer, and has much deeper taproots than red clover.

INCREASING ROOT EFFICIENCY

It has been well documented that plants with adequate fertility are more water-efficient. This is due to better root growth and better stomatal function and to simply having all the molecular parts available for biosynthesis when water is available.

To the farmer, none of this really matters. What does matter is that plants with adequate mineral fertility will grow a better root system and be more water-efficient. Research clearly backs up this theory (see Table 4.2 below).

How Levels of Nitrogen Affect Water Use Efficiency

These results are from a study done at Tifton, Georgia, showing improved water-use efficiency at different levels of nitrogen fertility.

SPECIES OF GRASS	WATER USED PER UNIT OF DRY MATTER		
	@50 LB N/ACRE	@100 LB N/ACRE	@200 LB N/ACRE
Common bermudagrass	9,268	4,437	3,294
Coastal bermudagrass	2,253	1,351	809
Suwanee bermudagrass	1,697	1,025	641
Pensacola bahia grass	2,970	1,829	1,180
Pangola grass	2,852	2,295	2,937

Table 4.2

On soils with low organic matter, conventional tillage, and no history of heavy organic amendments (such as manure or compost), routine soil tests available from any county extension agent or soil-testing laboratory are adequate for predicting fertilizer needs. To test your soil using these standard tests, simply visit your county extension office to obtain sampling instructions. However, with long-term no-till, cover cropping, and manure or compost application, new testing procedures can be better.

One test of benefit under these conditions is a Haney test, developed by Dr. Rick Haney, which attempts to more accurately predict the release of nitrogen, phosphate, and potassium (potash) from organic sources, usually ignored by ordinary soil tests. See Resources to find a few testing labs that offer the Haney test or an equivalent; I am sure that many others will begin to offer this useful test as well. This test is particularly useful in the beginning of a soil-building process to establish a "baseline" value for various soil health measurements.

Enlisting Mycorrhizal Fungi

One other factor involved in root efficiency is colonization with mycorrhizal fungi, which are mentioned many times in this book because they perform so many vital functions. The list of virtues of mycorrhizal fungi reads like a late-night infomercial: "But, wait! There's more!" One of the more important features is the ability of mycorrhizal fungi to increase the ability of plants to extract water from soil:

- Hyphae extend up to 24 inches past the root zone and increase the absorptive area up to 1,000 times that of an uncolonized root.

- The small diameter of the hyphae allows them to reach into pore spaces that root hairs cannot, helping the plant access a few more days of moisture between rain events.

- The glomalin secreted by hyphae enables them to penetrate compacted soil layers. Hyphae can also exert a mechanical force roughly ten times stronger than plant roots to break through mechanically resistant layers.

The benefits of inoculant on the drought tolerance of crops have been demonstrated by research, such as that conducted at the University of Wisconsin-River Falls by Hankes and Anderson (see References) in 2006. In this study, corn inoculated with mycorrhizal inoculant lived 28 percent longer when deprived of water than noninoculated corn. Thousands of research articles have been published on the drought-tolerance benefits conferred by mycorrhizal fungi, but despite these clear benefits, commercial inoculants have yet to be accepted by mainstream agriculture.

CHAPTER SUMMARY

- Subsoiling or any other deep tillage is not effective at improving root depth long-term.

- Lack of oxygen penetration is the real limiting factor to root depth.

- Soil oxygen is lowest during the summer and highest during the winter.

- The optimal time for growing a root to penetrate root-limiting layers is when soil temperatures are between 32°F (0°C) and 50°F (15°C).

- Deep-rooted cool season cover crops, like forage radishes, are ideal for breaking up root-limiting layers.

- Adequate soil mineral fertility improves root efficiency as well as plant water efficiency.

- Inoculation with mycorrhizal fungi can dramatically improve water uptake.

Losing crops to a drought is a disaster, but it pales in comparison to the loss a drought can wreak upon a pastured livestock operation. A crop loss is a one-year event, but drought on a ranch can result in the forced sale of breeding stock accumulated over decades of genetic selection and cause pasture degradation that reduces production for years to come. Crop prices usually rise during a drought, but livestock prices often crash as animals go to market prematurely because no one can feed or water them.

Learning to weather a drought may be one of the most valuable bodies of knowledge a rancher can accumulate.

PART II

PROVIDING FOR LIVESTOCK

5

ENSURING LIVESTOCK WATER SUPPLY

During the drought of 1980 our farm suffered severely. Our corn crop failed, and instead of harvesting grain we put many acres into a silo for later feeding, as the hay harvest was quite thin. Our pastures dried up, and some acres of corn were chopped and fed to the cows in the pastures every day. Most of our pastures, however, became useless even as crispy brown exercise lots, because there was no water in the ponds and the streams were dry.

We had to move most of our cows to the pasture with the "big pond" where we fed them chopped corn from the large failed field adjacent to it every day. My highlight of that summer was when a large group of neighbors got together with a 100-foot seine and salvaged the fish out of all the ponds that were almost dried up while we still could. We had a huge neighborhood fish fry that took our minds off the disaster taking place around us. Misery loves company.

That was the only time in my life that I have eaten fried snapping turtle. Our ponds served a purpose that year, but unfortunately it wasn't to provide livestock with drinking water.

HAVING WATER WHERE YOU NEED IT WHEN YOU NEED IT

During a drought, lack of drinking water is often a bigger problem for livestock than a lack of forage. Livestock are removed from pastures with adequate feed left in them because there is no drinking water after the pond or stream dries up. Ensuring reliable, adequate drinking water can help guarantee that it is always possible to utilize forage when it is available.

There is also deep value to having water sources strategically distributed throughout a pasture. Research trials done by Jim Gerrish in Missouri indicate that cattle poorly utilize grazing areas that are more than 900 feet from water. Having only a single water source in a pasture tends to concentrate grazing in areas close to the water source, resulting in overgrazing close to water and unutilized feed away from water. This is often aggravated by the fact that shade trees usually grow near water and by the common practice among ranchers of putting salt sources close to water sources.

Having multiple water sources around a pasture can greatly improve grazing distribution so that grazing pressure is uniform. You can dramatically improve pasture performance by preventing both overgrazing and undergrazing.

The first item of business is to develop a source of water of adequate quality and adequate quantity. The major sources are from surface water (like ponds and streams), subsurface water (springs and wells), and municipal or rural water systems intended for households.

SURFACE WATER

Surface water is a result of rainfall, which, as we all know, is variable, and it rains less in drier areas and drought years when we need water the most. Therefore, it is desirable to have a constant supply of water rather than the flood-or-dry cycles that come with rain-fed water sources.

POND DEVELOPMENT

Ponds are a highly desirable feature on the farm. They not only provide livestock water, but can also be used for fire-fighting, fishing, boating, swimming, irrigation, and simple enjoyment. A well-designed and well-maintained pond is a thing of beauty, and quite often the focal point of a farmstead. Unfortunately, most pasture ponds are nothing more than feces-laden mudholes that do a poor job of all the tasks listed above. The primary problem with most pasture ponds is unrestricted livestock access, as we'll discuss below.

Keep Water High on the Landscape

Water, of course, tries to move downslope. The longer this takes, the more water will infiltrate the soil. As discussed in chapter 2, one innovative system of providing water to a farm is keyline development, which involves sculpting the landscape to create small ponds at as high an elevation as possible. (See pages 34–35 for more information.)

Size Ponds to Fit Your Needs

Reliable water supply takes water storage. Bigger ponds can obviously supply more water than smaller ones. For a given volume of water, a deeper pond with a smaller surface area will lose less water to evaporation than a shallow pond with a large surface area. Round ponds lose less water to seepage than long thin ones, although a long thin shape may offer more habitat for fish and other aquatic life.

Keep Livestock out of the Pond

It is essential to keep animals out of water sources. When it gets hot, animals like to swim, which can greatly reduce heat stress. Allowing them unrestricted access to surface water, however, can cause a number of problems.

The first of these is the eventual failure of the water source, particularly ponds. A pond holds water because it is a hole in the ground surrounded by soil. If the banks of the pond are trampled by cattle or other livestock, erosion inevitably occurs and the soil from the banks soon ends up filling in the pond.

Second, animals standing in water tend to defecate, depositing parasites and disease organisms in the water. The fecal matter also is a source of nitrogen and phosphorus for the growth of blue-green algae that can be toxic.

Finally, it is also common for animals to get stuck in the mud or to fall through the ice when allowed free access to a pond. My father lost three heifers in one day when they fell through the ice on a pond with unrestricted access. That was disastrous not only because of the loss of three animals: it was impossible to safely remove the animals before they formed a soup of decaying flesh when spring thaw arrived, which ruined the pond as a source of drinking water for a year.

For these reasons, fencing a pond to exclude livestock is a wise investment. If reducing heat stress through swimming is necessary, it may be best to construct another pond for this purpose.

. . . But Provide Access to It

A limit-access structure can prevent livestock from damaging the banks of the pond while still allowing it to be used as a water source. Of course, livestock need at least limited access to a pond so they can drink from it. Therefore, you must devise a way to allow them to get water without entering it unrestricted. The ways to do this are limited only by your imagination.

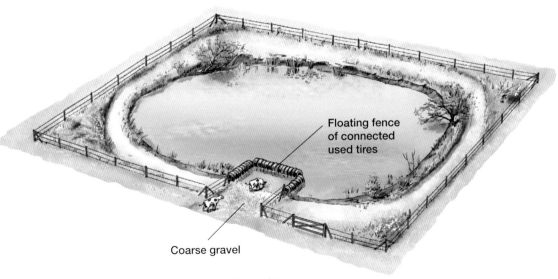

Floating fence of connected used tires

Coarse gravel

A fenced pond with a limit-access watering point

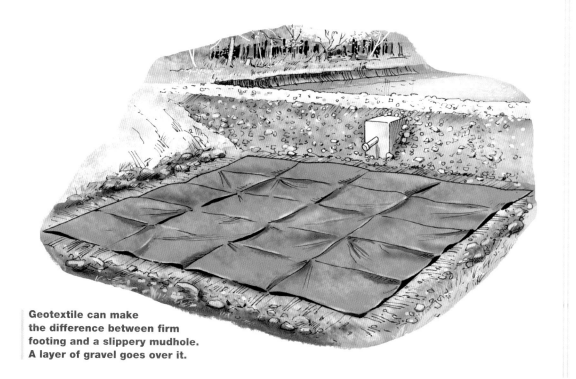

Geotextile can make the difference between firm footing and a slippery mudhole. A layer of gravel goes over it.

One way is to create a restricted-access area, with only a small portion of the pond bank accessible. Cover this area in a trample-proof surface, such as coarse gravel, preferably laid over geotextile fabric.

Often, hoof traffic pushes gravel into soil during wet periods, and then the gravel is gone. Geotextile is a sheet of tough woven plastic, similar to a woven plastic tarp, but it is water-permeable so it lets rain pass through while it stops the gravel from sinking into the underlying soil. For minimizing the effects of mud, it is a godsend — far superior to concrete pads, which provide a mud-free surface in one place but transfer water to the edge of the pad and worsen mud problems there.

Another way to get water *from* a pond without allowing access *into* it is to move the water outside the fenced area. You can do this by gravity, with a tube through the dam or with a siphon tube over the dam to any spot

Geotextile can also prolong the effective life of a gravel pad.

lower in elevation. Or you can pump the water through a hose to a tank anywhere outside the pond fence, even uphill.

Pumps can be powered by a variety of energy sources, including gasoline, wind,

Moving Water with Gravity

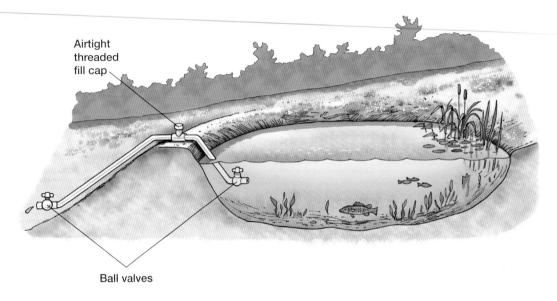

Airtight threaded fill cap

Ball valves

A pond siphon can be constructed from **PVC** pipe, which eliminates the need to create a hole in the dam to provide downhill water. The fill cap should be placed at the high point in the pipe curve over the dam. To create suction, close both valves, fill the pipe with water, replace the fill cap, then open the valves. As long as the exit pipe opening is below the level of the water in the pond, water should run out the pipe.

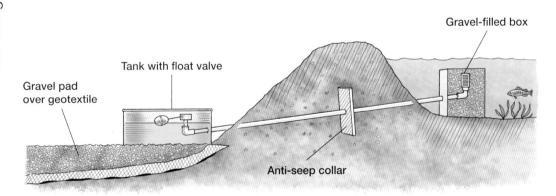

Gravel-filled box

Tank with float valve

Gravel pad over geotextile

Anti-seep collar

A pipe through a dam can provide water from a pond without having to allow livestock into the pond.

electricity, or solar-powered electric. As solar panels have become increasingly affordable, solar-powered pumps (often with a battery backup) have become the preferred way to move water uphill to a tank. You can set them up to pump constantly, with an overflow back to the water source, or to turn off when a float valve shuts off and backs pressure up against the pump.

STREAM DEVELOPMENT

Just as with ponds, allowing livestock unlimited access to streams can result in eroded streambanks and poor water quality due to fecal deposition and sediment load. Animals tend to congregate near streams for the close proximity to water, shade, and lush vegetation.

Some advisors advocate the complete exclusion of livestock from streams, but now progressive grazing managers are leaning toward allowing livestock to flash-graze streambanks for one or two days a season. This promotes the conversion of streamside vegetation into animal product. There is also accumulating evidence that occasional grazing stimulates better vegetation on the streambank. Willows and poplars, common streamside vegetation in my area, have very high-protein leaves and are particularly nutritious in late summer when many other forages are declining in value.

Therefore, it is desirable to devise a means of fencing that allows control of when and where animals are allowed into the stream. Pumping water out of the stream with a solar pump, water ram, or other means into a tank just uphill allows utilization of the water in the stream without degrading the stream. This also allows animals to get drinking water

with less exposure to mosquitoes and deer flies that often congregate around streams.

Just as with ponds, you can devise fencing to allow only partial access to streams. If the stream has much flow, however, these devices can be damaged during times of high stream flow. One solution is to construct a fence out of live willow trees. Willows can tolerate standing water and are easily propagated by inserting twigs into the mud. The twigs grow rapidly and form a very tough barrier over a couple of years (see next page).

In many areas, windmills are being replaced by solar-powered water pumps. While these lack the romance of windmills, they rarely break down and need very little maintenance.

Build a Willow Fence

You can create a living fence from willow cuttings by inserting twigs into moist soil in a row about one foot apart. This is most successful in the fall of the year, so roots can develop in the soil prior to leaf development the following spring. They tend to root readily from cuttings and grow rapidly.

Once the twigs develop a soft, pliable shoot of enough length, twist the first one into a loop and secure it with a cable tie. Insert the next twig in line through this loop and make it into a loop in turn, and so on down the line, so that the trees form an interlocking chain. Willow twigs thus held in place will fuse together at the point of contact created by the cable tie.

This fence will grow stronger each year, creating an impenetrable barrier to livestock. In addition, it will produce browse and shade, and a new crop of twigs each year for creation of new fences.

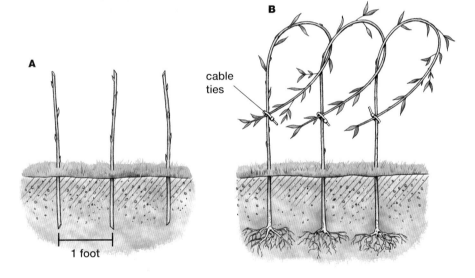

A
1 foot

B

cable ties

A willow fence is a natural, effective barrier that grows stronger every year. Insert dormant willow cuttings into moist soil, about 1 foot apart (A). Once the cuttings have rooted, bend one trunk into a loop and secure it with a cable tie. Pass the next twig through it to form an interlocking loop (B).

GUZZLERS

A guzzler is a structure that consists of a rainwater-collecting surface (usually corrugated roofing metal) that empties into a storage container (such as a large galvanized tank). Guzzlers depend on rainfall for recharge but can provide temporary drinking water for a short period of time in areas where water would otherwise be unavailable.

A good place to locate a guzzler is in a pasture area that is historically underutilized by grazing livestock, such as a hilltop far from water. A guzzler can provide both shade and water to attract animals there. It also makes complete sense to utilize the roofs of existing buildings to fill water containment structures. How often do you need to haul water to a barn for a confined animal? Wouldn't it be easier to have a tank right there that fills from the roof runoff every time it rains?

Metal roof with gutter

Storage tank

Drinking tank
with float valve

A guzzler can be used to create a temporary source of drinking water (and shade) that can attract grazing animals to areas of pastures too far from water for proper utilization.

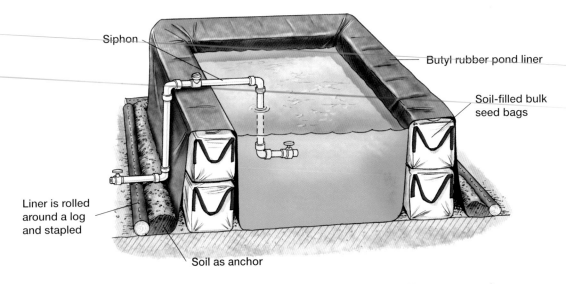

Siphon

Butyl rubber pond liner

Soil-filled bulk
seed bags

Liner is rolled
around a log
and stapled

Soil as anchor

An earthbag impoundment can be used to create an aboveground water reservoir
wherever desired, which can be filled with pumped water. Water can be siphoned
from the impoundment to fill water tanks placed at lower elevations.

Creating Cheap Water Containment Structures

Cost quickly becomes an obstacle to storing large quantities of water high on a landscape. I learned about a cheap way of building water containment structures in college. They say you learn more outside the classroom than you do inside it, and this was a good example.

My frat house hosted a beach party every year in the back yard, and we made our own beach and swimming pool. We had sand dumped in the yard and then filled a bunch of sand bags and stacked them in a ring. Then, we lined this structure with two layers of heavy-mil plastic. Presto! Instant swimming pool and surrounding beach! It worked great and went up quickly and cheaply.

The same basic design could work on your farm. And now there are more materials available that can do an even better job than our hand-filled sandbags. Used mini-bulk seed bags can hold several tons of soil or sand apiece and, once filled, are very stable due to their cubical shape. The bags are made of woven plastic that is very tough and virtually waterproof. You can also buy pond liners made of butyl rubber or EPDM, which are much more durable than plastic sheeting. These are widely available through pond supply stores.

The best thing about these structures is that you can place them on a hill and let gravity flow distribute water to tanks downhill. Use a siphon tube (see page 84) so you don't have to make a hole in the wall. Initially, the pond must be filled with water via any type of pump.

SUBSURFACE WATER

In many areas, surface water is available at very few locations, often farther than animals are willing to travel. Fortunately, groundwater is often available in many of these areas, such as the Great Plains. Since groundwater is the result of rainfall from eons previous, it is less dependent on recent rainfall than surface water and is often more reliable, though more expensive to develop. Development of groundwater can be a huge asset in ensuring adequate drinking water.

WELLS

In areas with a stable water table, wells are one of the most reliable forms of water supply. The problem with them, though, is that the water is underground and must be pumped aboveground. Traditionally, windmills were used to pump well water into above-ground tanks, but now the pump of choice is solar-powered, or photovoltaic. Windmills tend to need constant maintenance and are prone to breakdowns. Solar pumps need little maintenance, but they can operate only when the sun is shining (unless there is a battery backup) and thus need a larger storage tank than a windmill does. There are also devices called nose pumps that animals themselves power, though these are best used with very shallow wells.

Of course, the first item of business is to drill a well where you need it. There are various ingenious methods of drilling small wells, including one that uses a water hose inside a pipe to dig the well, illustrated below. You can also find many online videos showing this technique.

Drilling a Well with Water Power

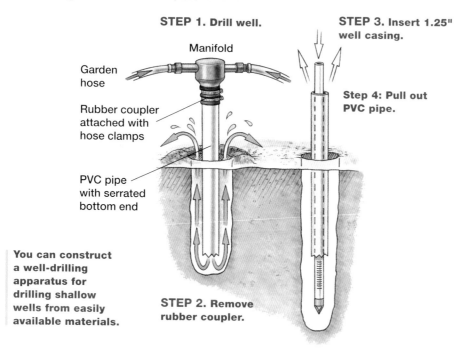

STEP 1. Drill well.

Manifold

Garden hose

Rubber coupler attached with hose clamps

PVC pipe with serrated bottom end

You can construct a well-drilling apparatus for drilling shallow wells from easily available materials.

STEP 2. Remove rubber coupler.

STEP 3. Insert 1.25" well casing.

Step 4: Pull out PVC pipe.

SPRING DEVELOPMENT

One pleasant side effect of improving water infiltration with better pasture management is that springs will begin to run again. Just keep in mind that if animals have unlimited access, springs become manure-filled mudholes. Spring developments can provide very high-quality water for very little cost compared to building a pond or digging a well.

A spring box is a structure designed to improve the cleanliness and reliability of spring water. To create one, dig a trench into the spring area, insert a perforated supply pipe, and backfill with coarse gravel to aid water movement into the pipe. The pipe is connected to a small water reservoir connected to a drinking tank. By forcing animals to drink out of the tank, rather than from the mudhole found surrounding most natural springs, the device keeps the water free of mud and manure. The water flow is also better when livestock aren't trampling the spring.

MOVING WATER WHERE IT'S NEEDED

Water can be transferred via open ditches or pipes to where it is needed. Open ditches can move water only downhill, while pipes can move water downhill or uphill.

The two most popular forms of pipe are PVC and high-density polyethylene (HDPE). PVC, the form usually used in buried water lines, comes in 20-foot lengths and is joined together with PVC solvent cement.

HDPE tubing comes in long coils and is tough enough to be laid along the surface because it does not degrade in sunlight, as does PVC, which should be buried. You can lay it out prior to the grazing season, then roll it up later and move it. This offers a great deal of flexibility.

Although HDPE tubing is not freeze-proof, it has enough flexibility that it will not burst if water freezes inside it. Most managers do drain it prior to freezing weather, however, just as a precaution.

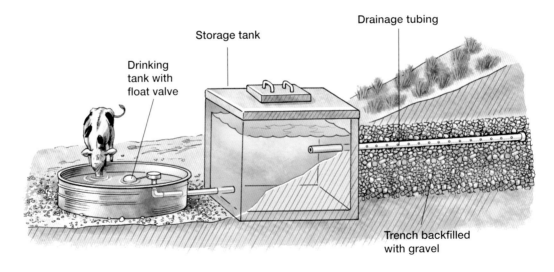

A spring box can create a reliable source of clean water from a hillside spring.

Laying it out along fence lines keeps animal traffic off the pipes, and the taller grass typically found along fences can shade the water lines and keep the water cooler. Burial of pipes also keeps them cooler; the water inside exposed black pipe can get very hot.

A Plasson quick coupler valve attached to high-density polyethylene pipe can be used to move water to a portable drinking tank where livestock are located, simplifying the logistics of rotational grazing.

ADDRESSING WATER QUALITY ISSUES

Not all water is good water. There are a number of pollutants that can drastically affect livestock health and can even be fatal. These issues are usually far easier to prevent than to fix once they have occurred.

NITRATE

Nitrate is a natural and necessary plant nutrient, and it comes from the bacterial breakdown of protein from plant residue, manure, dead animal bodies, or soil humus. It is also produced by microbial action upon nitrogen fertilizers. As desirable as nitrate is, it also has a dark side. If found in unusually high quantities in drinking water or feed, nitrate can produce a toxic reaction called **methemoglobin** in animals that consume it, in which the blood does not carry enough oxygen. This condition reduces animal performance and can be fatal. A symptom of this condition is brown blood.

Nitrate is most often an issue in well water in intensive agricultural areas that receive heavy fertilizer input, or in areas with heavy manure accumulation. Sandy soils are more subject to nitrate leaching into groundwater.

Preventing Problems

Prevention focuses on utilizing as much nitrate as possible in plant growth.

- Grass buffer strips and healthy vegetation around ponds and along streams can use nitrate before it gets into the water.

- Cover crops can use nitrate before it leaches into groundwater.

Do not apply fertilizer or manure close to ponds or streams, and do not let livestock stand in water that will be used for drinking.

Well water is more likely to have nitrate issues than surface water, as algae often use up nitrate in surface water. Routinely test your water. County extension agriculture offices offer water testing for a nominal fee, as do many private agricultural laboratories.

Resolving Problems

If water is high in nitrate, there are only two solutions: develop a different water source or remove the nitrate. One of the best ways to remove nitrate is to use it for its best use, as a plant nutrient. Putting high-nitrate well water into a pond and allowing water plants to grow allows the plants to naturally take up and metabolize the nitrate, and convert it into protein. After all, wetlands are "Nature's kidneys," and their role is to clean up water before it enters streams or groundwater. Municipalities create artificial wetlands for secondary sewage treatment at lower cost and with greater effectiveness than the standard mechanized treatment systems.

In the same way, you can construct a special pond for nitrate removal. Build it long, narrow, and shallow, to provide a lot of growing locations for aquatic plants. Design it in a serpentine pattern so that the water takes a long time to move through the pond from the inlet where well water is pumped in, to the outlet, where drinking water is taken out. Fill it with cattails, water lotus, duckweed, or other aquatic vegetation that can use nitrate as fertilizer. The water leaving the far end of the pond from the inlet should be much lower in nitrate.

Using high-nitrate water for irrigation can accomplish the same goal by benefitting terrestrial plants. This doesn't really solve the issue of converting high-nitrate water into drinking water, unless the water is somehow recovered from the tail end of a field (or from tile drains below the root zone) after the plants have used up the nitrate. It does convert the nitrate into an asset as nitrogen fertilizer.

BLUE-GREEN ALGAE

More correctly called **cyanobacteria,** blue-green algae grow in ponds, particularly those enriched by nitrogen or phosphorus from fertilizer or manure runoff. These algae can produce toxins that reduce livestock performance or can even be fatal. The algae are particularly toxic to canines, and dogs have been known to die very quickly after drinking contaminated water. The toxin also is potentially fatal to humans if ingested. Even washing hands or swimming in this water is discouraged.

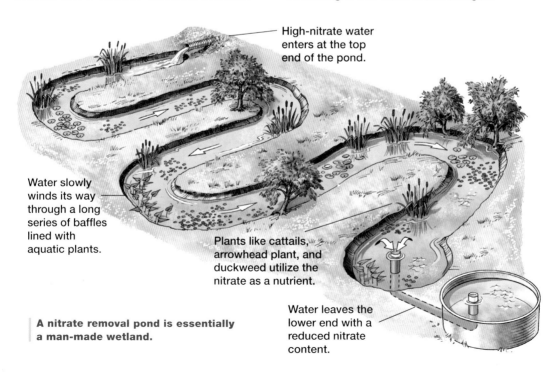

High-nitrate water enters at the top end of the pond.

Water slowly winds its way through a long series of baffles lined with aquatic plants.

Plants like cattails, arrowhead plant, and duckweed utilize the nitrate as a nutrient.

Water leaves the lower end with a reduced nitrate content.

A nitrate removal pond is essentially a man-made wetland.

One sign that toxic levels of blue-green algae are present is dead animals, such as mice, raccoons, or deer, near a pond. Blue-green algae is a visible condition, but only a water-testing laboratory can estimate the hazard level. Information on sampling and testing laboratories is available at your local county extension service.

Preventing Problems

One novel way to prevent the growth of blue-green algae is to place a bale of barley straw in the pond or tank. Barley straw contains a compound that is highly antagonistic to the growth of the cyanobacteria. Other straws, such as wheat straw, are not effective.

This method prevents algae growth but will not kill the algae or remove the toxin once growth has occurred. Therefore, it is necessary to put the bales into the water before you know you have a problem. As an estimate, you need one big round bale for a small pond, more bales for big ponds, and a little square bale for a stock tank.

In areas where ice becomes solid enough to support vehicle traffic, you can put bales on the ice in winter. In my area, we have to roll the bale down the bank and then push it along until it is in deep enough water to float. It may be prudent to anchor the bale to keep it in the desired location.

Another method of preventing blue-green algae is to introduce duckweed into the pond. Duckweed is a floating aquatic plant with one circular leaf about the size of a pencil eraser in diameter. It floats on the water surface and prevents sunlight from reaching the underlying blue-green algae, and also competes with it for dissolved nutrients.

Blue-green algae are more concentrated where the prevailing wind blows them. Where I live the summer wind blows from the south, so the blue-green algae are worst along the north shoreline. Having water access along the opposite shore helps reduce the risk of toxicity.

SALT

If a water source is contaminated with excess salt, there really is no economical way to reduce the sodium content. You must develop an alternative source of water.

SULFATES

Sulfates are often found in excess quantities in areas with deposits of gypsum rock (calcium sulfate). Sulfates in drinking water can cause severe diarrhea. If drinking water is high in sulfate, you must develop an alternative source of water, as there is no economically feasible way to remove it.

USING WATER TO HELP MOVE AND HANDLE LIVESTOCK

Many years ago I was visiting a large ranch in Arizona when I had one of those slap-your-forehead moments. This ranch had 11 paddocks, each about 7,000 acres in size. The landscape was solid with pinyon pines and junipers so that the visibility was only a few hundred feet at most. The rancher was explaining his rotational grazing system when I interrupted him to ask how in the world he moved the cattle. He looked at me as though I had just asked if brown cows gave chocolate milk, and he said the cows move themselves. I was quite baffled, and my expression showed it, so he explained.

Livestock require water every day and will travel to water on a predictable schedule. This can be put to use. The ranch placed their division fences so that a windmill stands along the division between two paddocks, surrounded by a catch pen. The pens are outfitted with what is called a **fish trap gate** using the principle of the inverted V found in fish traps. Features included:

- a set of double gates placed to form a funnel, with the narrow end pointed toward the interior of the pen

- a stop placed to leave a small gap at the end

- a spring or bungee cord that allows the gate to spread open but then snaps back into place

Fish trap gates can help move livestock from one paddock to another. Fish trap gates can also be placed leading into a catch pen surrounding the water source to capture animals for transport or processing, all without the usual rodeo and associated stress. I think of all the wasted time I spent whooping and hollering and chasing animals during my lifetime. Now I know a better way.

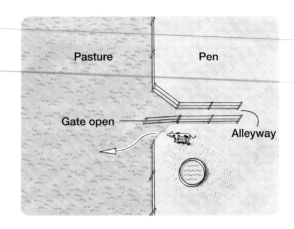

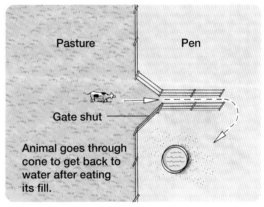

A fish trap gate can be used to move animals to a predetermined area without a person needing to be present.

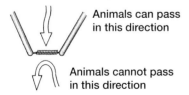

Animals can pass in this direction

Animals cannot pass in this direction

The opening of the fish trap gate is set slightly smaller than the shoulder width of an animal and can be pushed open by an animal moving through the funnel side of the gate. A spring snaps the gate back after an animal passes through. Animals cannot push back through in the opposite direction, so they are trapped on one side.

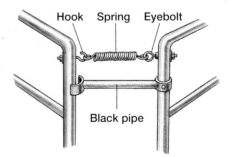

A metal pipe acts a stopper to ensure that the gate is left open enough for an animal to fit its head through.

The Drought-Resilient Farm

CHAPTER SUMMARY

- In a drought, drinking water is often more critical for a livestock operation than feed is.

- Retain water as high on the landscape as possible so it can be distributed by gravity.

- Fence your ponds to exclude livestock. This prolongs pond life, maintains animal safety, and provides cleaner drinking water by preventing defecation in the pond.

- Access water from a fenced pond using siphons, tubes through the dam, uphill solar pumping, or limited-access fencing.

- Fence streams from livestock to prevent stream degradation. Allow animals to drink from limited access points, or pump water uphill to a tank.

- Use guzzlers to provide occasional water at points far from natural water sources, such as hilltops. This can greatly aid grazing distribution.

- Construct quick and cheap above-ground water impoundments using earth-filled bags lined with a butyl rubber pond liner.

- Dig wells by traditional methods, or with an innovative water jet device — a high-pressure water hose inside a pipe.

- Develop springs with spring boxes to improve the cleanliness and reliability of your water supply.

- Manage nitrate levels in drinking water by excluding animals from the water supply, maintaining a vegetative buffer around ponds and streams, planting cover crops to utilize off-season nitrate, and carefully managing fertilizer.

- Nitrate can be removed from contaminated water by growing plants.

- Blue-green algae can be prevented by keeping manure and fertilizer out of water.

- Barley straw appears to be toxic to blue-green algae, and a bale of barley straw floating in the water is effective for prevention.

- Test for salt and sulfate content and develop alternate water supplies, if necessary.

- Water access can be used as a means for moving or capturing livestock with fish trap gates.

6

CREATING DROUGHT-TOLERANT PASTURES

The top three rules in pasture management are as follows:

1. NEVER OVERGRAZE.

2. NEVER OVERGRAZE.

3. NEVER OVERGRAZE.

This sounds easy enough. Just make sure your animal numbers match up with pasture growth. But what happens in a drought, when the pasture growth slows or stops, and animals are still eating the same amount as usual?

Perhaps your answer is to buy hay. Have you ever noticed what happens to hay prices during a drought? Buying hay out of necessity during a drought is a very rapid route to bankruptcy.

A far better option is to have a pasture that will keep growing in a drought. How in the world can you have a pasture that will grow through a drought? Read on.

PASTURES WILL MAKE THE MOST OF INPUTS

It is common knowledge that a pasture with adequate fertility and proper pH will produce more in both good years and bad, the benefit being greater in bad years. At the same time, it is also very common for pastures to be nutrient deficient.

Farmers often see pastures as a low return on investment acre for fertilizer input, compared to crop fields. I cannot understand this view. One application of fertilizer to a pasture will recycle back through manure and urine over and over, unlike a grain or hay crop, in which the nutrients are removed every year. The return on a pound of phosphorus used 20 times in a pasture over 20 years is far better than the return on a pound of phosphorus used to grow corn grain that gets used once and is then exported in the grain.

SOIL TESTS BOTH SHALLOW AND DEEP

Soil testing can identify any limiting fertility factors. I suggest taking a shallow — 2 inches or less — sample, along with the standard 6-inch depth. Pastures receiving nitrogen fertilization often develop a layer of surface acidity that prevents seedling development of many legumes, such as clover. This layer is often undetected when its soil is blended with all the underlying soil in a standard test. If there are lime recommendations based on this surface layer, be sure to divide the amount by three, as standard lime recommendations assume that we are neutralizing a 6-inch layer, not a 2-inch layer.

I recommend sampling multiple locations throughout a pasture because fertility is usually quite variable. Grazing tends to redistribute nutrients toward water and shade. These areas often test quite high, while other areas test quite low. For this reason, it's best to sample high animal traffic areas separately and manage them separately. This is also why you should distribute shade and water sources throughout the pasture, rather than concentrating them in a few points, for long-term pasture productivity and animal performance. Locating shade or watering points in low-fertility areas can help remedy poor nutrient distribution. Pastures are good places to sample with a Haney test (see page 74), since the typical active biology in a pasture makes it more efficient at cycling organic sources of nutrients than cropland is.

SPECIES SELECTION

There are very real differences in the ability of different pasture plants both to survive and to produce during drought. Note also that there is a very real difference between drought tolerance and the ability to produce during a drought. For example, smooth bromegrass is one of the more drought-tolerant cool-season grasses planted throughout much of the Midwest. Its survival mechanism is that it

Smooth bromegrass in bloom

simply goes dormant during the summer and doesn't even try to grow. Obviously, while this may save the expense of having to reseed a pasture, it doesn't do much good for supplying livestock grazing during dry summers.

Drought production comes from several plant mechanisms, described as follows.

CARBON-FIXING PATHWAYS

All plants live by photosynthesis, which is the fixing of carbon dioxide into sugar. Only in the 1960s was it discovered that not all plants photosynthesize in the same manner; there are multiple pathways by which plants do this. Three main types have been discovered so far, with variations within their mechanisms.

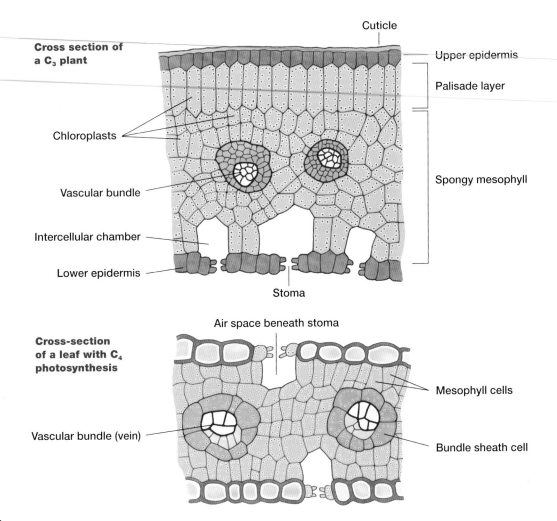

Cross section of a C₃ plant

- Cuticle
- Upper epidermis
- Palisade layer
- Chloroplasts
- Vascular bundle
- Intercellular chamber
- Lower epidermis
- Stoma
- Spongy mesophyll

Cross-section of a leaf with C₄ photosynthesis

- Air space beneath stoma
- Vascular bundle (vein)
- Mesophyll cells
- Bundle sheath cell

C₃ Pathway

The first type of photosynthesis to be discovered has been titled C₃, referring to the number of carbon atoms in the first product of photosynthesis in the pathway. This is the method used by cool-season grasses and the vast majority of trees, shrubs, and broadleaf plants (see list on facing page), and it has the poorest water-use efficiency of the three strategies.

Here's why. The photosynthesis process consists of two parts: the light reactions, which require light, and the dark reactions, which do not require light. The light reactions involve the splitting of water into hydrogen and oxygen. The hydrogen is used in the dark reactions, while the oxygen is a waste product, as we will see later. The dark reactions involve the combining of carbon dioxide and hydrogen (from the light reactions) into glucose.

In C₃ plants, the light and dark reactions take place in the same cells, the mesophyll cells. This is a convenient arrangement, allowing easy transport of the hydrogen from light reactions to dark reactions. When it's sunny and warm out, however, and the light reactions are occurring very rapidly, so much oxygen is accumulating in the cell that the enzyme responsible for grabbing CO_2 will grab oxygen by mistake. The resulting process ends up *not* producing glucose.

This "monkey wrench in the gears" phenomenon is called **photorespiration**. Photorespiration acts to put a ceiling on plant productivity during high temperatures and intense sunlight. On hot sunny days, plants with a C_3 photosynthesis can utilize only about 50 percent of the available sunlight and may actually lose biomass when leaf temperatures exceed 86°F (30°C). As long as plants have adequate water, evaporative cooling from transpiration can keep leaf temperatures low enough despite high air temperatures. Heat and drought combined, however, can cause C_3 plants to simply stop growing. That is why cool-season grasses, like tall fescue, orchardgrass, and smooth bromegrass, grow very little in a typical Midwestern summer.

C_4 Pathway

Some plants have evolved a photosynthetic method to circumvent photorespiration and thus be more productive in hot, sunny weather. This method is referred to as C_4, again because the first product has four carbon atoms in its structure. Examples of C_4 plants include warm-season grasses, such as corn, sorghum, sugar cane, and bermudagrass, as well as pigweeds and some desert shrubs, like saltbush.

Two mechanisms help these plants avoid photorespiration. The first is the separation of the light reactions and the dark reactions into different cells. The light reactions take place near the surface of the leaf, in the spongy mesophyll cells, where the close proximity to the stomata makes it easier for the oxygen to escape. The dark reactions take place in a set of cells close to vascular tissue (the bundle

sheath cells), where the glucose can be moved quickly to other parts of the plant. C_4 plants also have a mechanism that pumps CO_2 from the outer part of the leaf into the inner layer where the dark reactions take place.

Because there is no photorespiration, C_4 plants grow very rapidly in hot, sunny conditions and actually reach peak productivity around 90°F (32.2°C) with adequate moisture. (In contrast, most C_3 plants are most productive at about 77°F [25°C].) They also can utilize more sunlight than C_3 plants can. Due to these advantages, on a hot sunny day a C_4

C_3 and C_4

EXAMPLES OF C_3 GRASSES	EXAMPLES OF C_4 GRASSES
Kentucky bluegrass	Bermudagrass
Tall fescue	Bahia grass
Smooth bromegrass	Dallis grass
Orchardgrass	Big and little bluestem
Reed canarygrass	Indiangrass
Perennial ryegrass	Corn
Wheat	Sorghum
Rye	Switchgrass
Oats	Eastern gamagrass
Barley	Gramas (blue, black, hairy, sideoats)

EXAMPLES OF C_3 BROADLEAVES	EXAMPLES OF C_4 BROADLEAVES
All legumes	Pigweeds
All brassicas	Lambsquarters
Sunflowers	Saltbushes

plant can produce three times the daily biomass of a C_3 plant.

On the other hand, C_4 plants are very unproductive below a temperature of 55°F (12.8°C) and in temperate regions have very short growing seasons. C_4 plants are more efficient in use of water and nitrogen but are also less digestible (at similar stages of maturity) by ruminants. Residue of mature C_4 plants tends to be extremely indigestible and very low in protein.

CAM Pathway

The final photosynthetic pathway, and the one with by far the best drought tolerance, is the Crassulacean Acid Metabolism (CAM) pathway, found in succulent plants like cacti, yuccas, and pineapple. These plants close their stomata during the day to prevent water loss, and they open them at night. Since stomata are what plants use to "inhale" CO_2 for photosynthesis, this method requires the plants to store the CO_2 taken in at night for use in photosynthesis the following day. The CO_2 is stored by attaching it to a compound called crassulacean acid, hence the name.

Because this process requires only a fraction of the water needed by other pathways, many farmers have explored using CAM plants, such as spineless prickly pear cactus, as a pasture resource. The problem is that these species grow very slowly, even under ideal conditions. CAM plants take about 4 years to produce as much growth as 1 year of grass would under the same conditions. As a result, it is much more economical to grow grass and simply adjust the stocking rate to a level that leaves enough ungrazed grass as an emergency reserve.

This is not to say that CAM plants have no value in a drought survival strategy. In many pastures in arid regions, there are patches of cacti that can be used by livestock in an emergency, once the spines are removed. Two common methods of removing cacti spines are to flame them and to fine-grind the entire plant into a pulp. Both methods are expensive and labor-intensive, but because cacti removal is often a desired goal anyway it can turn a spiny nuisance into an asset. Many cacti have very high digestibility but tend to be very low in fiber and protein. They also tend to cause scouring (diarrhea), so supplementation or dilution with other feeds is recommended to supply fiber and protein and dilute the diarrhea-causing compounds.

Summarizing the Differences

Basing your summer pasture resources on C_4 grasses will greatly reduce the impact of summer droughts on pasture production. To produce a pound of dry matter in the summer, a C_4 grass requires only about one third of the water required by a C_3 species. Despite this obvious advantage, most pastures in the Midwest and northeastern United States are based on C_3 species because when drought stress is not a concern the longer growing season of C_3 grasses tends to result in higher season-long productivity than that of C_4 grasses.

How C_4 and C_3 Plants Compare in Water Use Efficiency

This table illustrates the superior water use efficiency of C_4 species compared to C_3 species, in grams of plant produced per kg of water used.

	C_4 SPECIES	C_3 SPECIES
Broadleaf plants	3.44	1.59
Grasses	3.14	1.49

Table 6.1

Progressive pasture managers, however, have realized that it is quite possible to interseed warm-season annual plants into C_3 pastures to increase summer productivity and add to total seasonal yield. An example of this is to interseed teff grass or an improved variety of crabgrass, such as Red River or Quik N Big, into an existing stand of fescue or orchardgrass. Similarly, in areas dominated by C_4 species it is common to interseed cool-season annuals into warm-season sods, such as interseeding annual ryegrass into bermudagrass.

Another barrier to increased adoption of these more efficient C_4 species is the perception that they have a very long establishment period, often taking 3 years to reach full productivity. We now know that this long establishment period is due to these species' high dependence on mycorrhizal fungi, which are almost always absent from any field that has been formerly cropped. Inoculating the seed of native C_4 species with spores of mycorrhizal fungi (see Resources) at planting can result in much faster establishment of these valuable species as well as better long-term productivity. (See pages 24–25 for fuller discussion of mycorrhizal fungi.)

Smooth bromegrass by itself (top), and with interseeded summer annuals (bottom)

Interseeding Annual Warm-Season Species into Cool-Season Grass Sods: A Case Study

One enterprising farmer I know has for several years interseeded warm-season annual pastures into his tall fescue and smooth bromegrass sods after they are hayed or grazed off in June. This practice began when he drilled some leftover cover crop seed into the fescue just to clean out his drill. (It is better to feed the soil, after all, than the mice in the seed shed.) He liked the results so much that he continued the practice on his own land, then recently expanded it to his rented pastures. His current mixture contains a brown midrib grazing corn, a brown midrib sorghum-sudan, cowpeas, sunn hemp, forage soybeans, and millets.

In the fall of 2016, he had the NRCS rangeland specialist come out and measure the biomass of the interseeded area compared to the control area. Over several samples there was an average of roughly 3,000 pounds per acre of additional nutritious pasture in the sod where he interseeded for the first time. But here is the amazing part: where he had interseeded for the fourth year in a row, he took two samples; one had 5.5 tons of standing biomass, and the other had 8 tons of standing biomass. It seems that this ground is increasing in productivity every year he does this; his hay yields the following year are higher in the portions of the field where he has interseeded, by over a third of a ton per acre.

Apparently, growing several additional tons of biomass, producing nitrogen from the legumes (cowpeas, soybeans, sunn hemp), and converting it into several tons of manure per acre benefits the soil. Who would have thought? The fescue looks green and healthy beneath that jungle, appearing to benefit from this practice rather than suffer from the competition. In fact, a visual comparison would say his fescue looks better than any other in his area. Perhaps the shade of the overstory plants keeps the fescue cooler in the heat of summer. Or perhaps there really is something to the idea that additional biological activity makes land more productive.

ROOT DEPTH

Including deep-rooted plants in pastures can result in the accession of water from deep in the soil profile, which may be unavailable to typical pasture grasses. Pastures in the northeastern US often become dominated by Kentucky bluegrass and white clover, because those species are so tolerant of heavy continuous grazing. Unfortunately, they are also very shallow rooted and susceptible to drought. Species like chicory, alfalfa, and plantain can reach several feet deep into the soil, and they do more than obtain water for themselves. It has been found that these plants bring up an excess of moisture during the day and then release small amounts of it into the soil near the surface at night, effectively irrigating their neighbors to a small degree. Eastern gamagrass and reed canarygrass are not only quite deep rooted, but have a tissue in their roots called **aerenchyma** that allows them to grow into a water table.

Obviously, deeper-rooted plants can access water longer than shallow-rooted plants. The roots of mesquite trees, for example, may reach 100 feet deep in some situations.

Deep-rooted forage crops: Alfalfa, big bluestem, chicory, eastern gamagrass, Indiangrass, reed canarygrass, switchgrass

Moderately rooted: Smooth bromegrass, bermudagrass, red clover, tall fescue

Shallow-rooted: Kentucky bluegrass, white clover, orchardgrass, perennial ryegrass

PUBESCENCE

Leaves that are hairy act as their own wind-break. For example, pubescent wheatgrass with hairy leaves is slightly more drought-tolerant than the closely related intermediate wheatgrass with smooth leaves. The trade-off of hairy leaves is that they also create their own shade, so they tend to be slightly less productive under well-watered conditions.

LEAF FOLDING

Plants' ability to fold or roll their leaves can dramatically reduce water loss. For example, the prairie legume catclaw sensitive briar folds its leaves when touched by a neighboring plant. It also folds up in a strong wind to reduce water loss. Sunn hemp and Illinois bundleflower fold up their leaves at night to reduce water loss (this makes a lot of sense; the sun is not out for photosynthesis anyway). Some grasses, such as sorghum, readily roll their leaves under drought stress.

Catclaw sensitive briar (*Mimosa nuttallii*) closes its leaflets when touched.

WAXY CUTICLE

Some plants, such as sorghum, feature a layer of wax over the leaf surface that reduces water loss through the leaves — one reason sorghum is more drought tolerant than corn. This adaptation is often found in tropical and desert plants.

AVOID OVERGRAZING WITH A HEALTHY STOCKING RATE

The best insurance against reduced pasture productivity during a drought is to have healthy plants with deep root systems, adequate carbohydrate reserves, and a healthy layer of mulch to promote infiltration and reduce evaporation. Unfortunately, the vast majority of pastures worldwide are currently in an overgrazed state with weakened root systems. It is critically important to have a stocking rate that is conservative enough during "average years" that stocking rate reductions are unnecessary in bad years. Table 6.2 illustrates the impact of grazing on carbohydrate reserves.

As you can see, grazing even a little too much can dramatically reduce plant vigor, so it is very important to establish a proper long-term stocking rate. At first glance, it would seem that a rate that removes 40 percent of the foliage would be appropriate. But bear in mind that the 40 percent is an average and that in some years the removal will be greater than that, therefore causing considerable damage. The proper stocking rate is sufficiently conservative to ensure that no more than 40 percent is removed in the *worst* years.

How Grazing Affects Carbohydrate Reserves

Removal of more than 40% of the leaf area results in a loss of root growth; this loss becomes severe once defoliation exceeds 50%. This research was the basis for the "Take Half and Leave Half" philosophy of range management, although clearly that should be modified to say, "Take no more than 40% and leave at least 60%."

% LEAF REMOVAL	% NON-GROWING ROOTS		
	RhG	SB	KB
10	0	0	0
20	0	0	0
30	0	0	0
40	0	0	0
50	2	13	38
60	50	36	54
70	78	76	77
80	100	81	91
90	100	100	100

RhG = rhodesgrass, SB = smooth bromegrass, KB = Kentucky bluegrass

Table 6.2

Obviously, income maximization is important, but so is maintenance of the resource base. It is a persistent myth that income maximization is achieved through overgrazing. I have visited many ranches where the owner apologized for his obviously overgrazed pastures, then rationalized it by saying something along the lines of "I know my pastures are overgrazed, but I just have to make all the money I can off this grass." The data below, from the Fort Hays Experiment Station of Kansas State University, demonstrates how the best long-term stocking rate for grass vigor was also the one that produced the most income through better livestock performance. There is no real economic incentive to overgraze. You make more money from conservative grazing than you do from exploitative grazing.

Clearly, the heavy stocking rates resulted in the least profit in each scenario. Abusive grazing causes poor animal performance, which results in poor profitability. The "I have to overgraze because I need to make all the money I can" excuse is not valid at all.

What is meant by a "heavy" or "moderate" or "light" stocking rate varies by region and rainfall and even soil type. A stocking rate considered light on a pasture in eastern Texas with 45 inches of annual rainfall may be disastrously heavy in western Texas with 15 inches of rainfall. The best way to determine the optimal long-term stocking rate for a pasture is to use the information and tools available at the local Natural Resource Conservation Service in the county USDA office. Local NRCS staff can calculate an accurate beginning stocking rate based on local average rainfall, pasture condition, and soil types present in the pasture. This service is free of charge and can be quite valuable, particularly for the novice rancher.

How Stocking Rate Affects Cattle Gain and Profit

This data, comparing the profitability of different stocking rates, is from the Kansas State University Experiment Station at Hays, Kansas.

STOCKING RATE	$/HEAD	$/ACRE
Light (5.0 acres per head)	$26.04	$5.21
Moderate (3.4 acres per head)	$16.91	$4.97
Heavy (2.0 acres per head)	-$7.31	-$3.65

This study was updated with current prices by Keith Harmoney of Kansas State for the period 2006–2010. I thought the original data interesting because the light stocking rate was the most profitable, and the heavy actually lost money. The Harmoney figures probably reflect more current conditions, but the original illustrates my point that proper grazing is more profitable than abusive grazing.

STOCKING RATE	$/HEAD	$/ACRE
Light (5.0 acres per head)	$91.70	$18.34
Moderate (3.4 acres per head)	$75.64	$21.63
Heavy (2.0 acres per head)	$24.95	$12.48

Tables 6.3 and 6.4

Wood Chip Mulch

Some progressive producers are using wood chips to mulch their pastures, having found that once they begin to create a more biologically active pasture, their grass residue rots away prematurely. While the more biologically active system is quite desirable, the soil cover disappears too quickly and leaves the soil bare and subject to evaporation, sunbaking, and erosion. Wood chips, because of their very low protein content, decay much more slowly, and provide a persistent residue to hold up cattle hooves in wet weather and prevent compaction. A mulched pasture also offers incredible infiltration rates as well as low rates of evaporation, contributing to a much more drought-tolerant situation.

It is when pastures are muddy, however, that wood chips offer their greatest

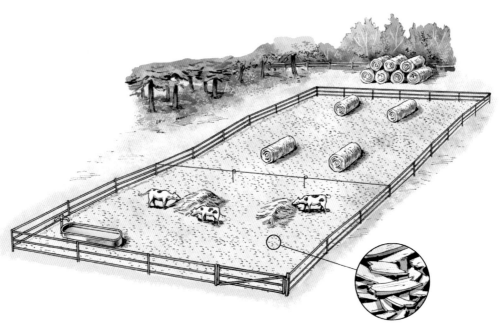

When pastures become saturated and unable to support livestock without pugging, a wood-chip pad can be combined with spaced bale feeding to create a mud-free area.

value. A pad of wood chips used as a feeding area eliminates the pasture damage done by hoof traffic in soggy conditions, especially during spring thaw. A pasture that is trampled during wet weather can decline greatly in productivity, often for a full growing season. Concentrating the feeding on a chip pad can prevent that damage. The blend of chips and manure can then be spread over the rest of the pasture the following season to enhance fertility and mulch the surface.

Research has shown significant advantages of using a wood chip winter feeding pad, including reduced winter hay needs and improved animal performance. Wood chip pads can be particularly effective in reducing diseases caused by poor sanitation and muddy conditions, such as calf scours. Wood chips can often be obtained at little or no cost from tree-trimming services, utility companies, and city and county maintenance operations, since their alternative is usually to pay a tipping fee at a landfill. The wood chips pay for themselves during the winter, and their mulch value comes as a free bonus later.

An alternative to a thick pad of chips that needs to be mechanically spread later is to apply a thin layer of chips over a larger area and keep the animals moving to prevent pasture **pugging** during wet periods. This enables more area to be covered by the same amount of chips, and it eliminates the labor and equipment cost of mechanical spreading.

It is often said that wood chips can reduce nitrogen availability in the soil due to their extremely high carbon-nitrogen ratio. This is true for sawdust, but wood chips offer far less surface area that can contact soil than sawdust does, so any nitrogen-depleting effect is very slow, and the effect at any given time is minor.

Properly grazed pastures (top) have much deeper root systems than overgrazed pastures (bottom).

Also, a well-managed pasture will have a much deeper root system than an overgrazed pasture, capable of extracting more water when needed. Pastures with adequate stubble will have much better infiltration of rainfall when rain does occur, as Table 6.5 illustrates.

Importance of Growing Point Location

It is also important to understand that the ability of grass to tolerate extreme grazing pressure changes throughout the life cycle of the grass. Much of this change in grazing tolerance is due to the location of the **growing points,** or the places on the plant from which leaves originate, which change throughout the season. There are three stages of a grass life cycle:

The vegetative stage is from green-up until the beginning of seedhead formation. During this stage, the grass has mostly leaves above ground, and the growing points are below the reach of grazing animals. Also during this stage, defoliation has a small impact on growth, and only if the defoliation is severe.

The reproductive stage is from the beginning of seedhead formation until ripe seed. During this stage, the growing points are elevated, and the grass becomes very sensitive to severe defoliation.

Dormancy is when the plant is not growing in winter. During this stage, the grass is extremely tolerant of defoliation and can be completely grazed off without harm to the grass; however, the quality during dormancy is usually poor for most grass species.

The timing of each of these stages differs between cool-season and warm-season grasses. Cool-season grasses begin their

How Grazing Intensity Affects Rain Infiltration

	LIGHT GRAZING	MODERATE GRAZING	HEAVY GRAZING
Rain infiltration, inches per hour	1.58	1.19	0.73
Vegetation left after grazing, pounds per acre	1,271	881	564

Table 6.5

vegetative growth cycle in fall, go dormant in winter (in areas with a winter), resume vegetative growth in early spring, and are reproductive in late spring. Warm-season grasses begin vegetative growth in late spring or early summer, are reproductive in late summer and early fall, and go dormant around time of first frost.

In my area of Kansas, cool-season grasses are reproductive between mid-May and mid-June and are thus most sensitive to extreme defoliation during this period. Warm-season grasses are reproductive from late July through late September and are thus sensitive to severe defoliation during this time. Most droughts in my area occur in late summer, and this coincides with the time when warm-season grasses are most susceptible to severe defoliation.

Given the information in Tables 6.6 and 6.7, it's clear that we should avoid close defoliation of cool-season grasses in late spring, when they are trying to produce a seedhead. This is usually easy enough, as this period typically coincides with the most productive time of the year on perennial pastures. It can also be seen that warm-season grasses benefit from a rest in late summer, when they are producing seedheads. This is tougher to accomplish because few, if any, perennial pastures are productive during that time period. In this situation, cover crops planted into the stubble of a summer-harvested crop, like wheat, oats, or peas, become invaluable for grazing, to allow rest for perennial pastures. Cover crops for pasture are further discussed in chapter 7.

When Severe Defoliation Most Affects a Cool-Season Grass

This table shows when weekly clipping of wheatgrass in Colorado to a 1-inch stubble height is most detrimental and threatens plant survival.

PERIOD CLIPPED	PLANT SURVIVAL
Apr 15–May 7	100%
May 1–May 22	25%
May 15–June 7	50%
June 1–June 22	55%
Every 2 weeks, Sept 15–Nov 1	100%

Table 6.6

When Severe Defoliation Most Affects a Warm-Season Grass

This table shows how mowing date affects the carbohydrate reserves of big bluestem in Kansas.

MOWING DATE	% CARBOHYDRATE IN ROOTS
Unmowed	7.0
June 1	8.25
July 1	8.5
Aug 1	6.7
Sept 1	4.4

Table 6.7

What exactly constitutes severe defoliation varies by species as well. Those with low growing points can tolerate grazing better than upright plants with elevated growing points.

Plants with low growing points include white clover, perennial ryegrass, fescue, blue grama, buffalograss, bermudagrass, Kentucky bluegrass, and tall fescue.

Plants with elevated growing points include smooth bromegrass, switchgrass, and Indiangrass.

Others: Orchardgrass is intermediate. Eastern gamagrass requires very tall stubble heights.

A simple drive through the countryside at the end of the growing season will reveal that the minimum stubble heights recommended in Table 6.8 are maintained in very few pastures. The sad reality is that overgrazing is the status quo. At the same time, this should also give hope that the productivity and drought tolerance of most of our pastures could be dramatically improved without any additional cost, just by pulling animals off before grazing below these minimum stubble heights. The 3 to 6 inches left over in the pastures is not wasted; it can be grazed off after the onset of dormancy without harm to the grass. Methods to improve the utilization of dormant grass are discussed in chapter 7. Making dormant grass more valuable is an important part of improving pasture management. People will manage to leave more stubble, if the stubble they leave to be grazed in the dormant season becomes more valuable than the grass they grazed during the growing season.

Minimum Recommended Grazing Height during Growing Season

Do not graze below this minimum height during growing season.

SPECIES	MINIMUM HEIGHT
Kentucky bluegrass	3"
Smooth bromegrass	6"
Buffalograss	3"
White clover	3"
Tall fescue	4"
Eastern gamagrass	12"
Blue grama	3"
Indiangrass, big bluestem	6"
Orchardgrass	4"
Perennial ryegrass	3"
Switchgrass	6"

Table 6.8

CHAPTER SUMMARY

- Plants with a C_4 physiology are more water-efficient than plants with a C_3 physiology.

- Characteristics that impart drought tolerance include deeper root depth, pubescence, leaf folding or rolling, and a waxy cuticle on the leaf surface.

- Grazing intensity that removes more than 40 percent of the plant topgrowth results in reduced root growth.

- Heavy stocking rates are less profitable than moderate stocking rates because individual animal performance is reduced.

- The local NRCS is a good place to determine a good beginning stocking rate for a pasture based on soil types and local conditions.

- Maintaining adequate residue cover makes a pasture more drought-tolerant due to better rainfall infiltration and reduced evaporation.

- Grasses are least tolerant of heavy grazing during their reproductive stage, most tolerant after dormancy, and moderately tolerant during the vegetative stage. Forage quality is highest during the vegetative stage and poorest after dormancy.

- The minimum recommended stubble height varies by species. Species with low, protected growing points tend to be more tolerant of close grazing than species with elevated growing points.

- The single biggest key to improving drought tolerance of a pasture is proper grazing pressure.

7

PROVIDING EMERGENCY FORAGE IN A DROUGHT

In the depths of the 2012 drought, the county extension agriculture offices and USDA offices conducted drought advice seminars for ranchers all across the country. After attending a few, I concluded: what a waste of time. Basically, each 2-hour seminar told us either to wean early (reducing this year's paycheck) or to cut our herd size (reducing our paycheck for years to come).

I shouldn't be so hard on them. The advice to destock is every bit as valid as telling an overweight person to eat less. The point is that it is simply not useful, because the advice is not taken, any more than the advice to eat less. We know it is dry, and we know we need to destock. But we don't do it. We keep the cows around until the last bite of feed is eaten, the pastures are bare and dusty, and the hay is gone, hoping that a rain is right around the corner. Those cows are our livelihood, and parting with them is extremely tough. Advice to the contrary is just not welcome, and thus it is usually ignored until it is too late.

The intention of this chapter is to offer practical alternatives to destocking. Destocking is a quick road to economic hardship — although if no other measures are taken, it will be far preferable to *not* destocking, as overgrazing in a drought can easily reduce the productive capacity of the land for generations to come.

FAILED CROPS

One of the most obvious consequences of a drought is that field crops will probably fail. Even so, most field crops can make excellent pasture, hay, or silage. Corn, soybeans, wheat, sorghum, and alfalfa can all be grazed during a drought, as well as mechanically harvested. I strongly recommend grazing over mechanical harvest, if possible, particularly if the crop is on land you own.

SILAGE PROS AND CONS

There is an assumption that mechanical harvest produces the most feed value per acre, particularly with silage. This is not necessarily true. While the very small amount of plant

The Long View

As I have gotten older, I have come to enjoy areas of study that I used to find mind-numbingly boring, perhaps because I now see how relevant they are. Archaeology and history are two of those fields of study. I have lived long enough to see too many people repeat the mistakes of the past because they do not know history.

History tells us that if we try to maintain livestock numbers through a period of reduced forage growth without other measures, taking it can result in long-term land ruination. Hopefully, by the time the next drought rolls around you will have already managed your soil to have better infiltration (chapter 2), better water-holding capacity (chapter 3), and deeper-rooted and more water-efficient plants (chapter 4). You will have managed your pastures so that they are composed of drought-tolerant species and so that those plants are vigorous with healthy roots (chapter 6).

If you follow the advice of the previous chapters, and if you are fortunate, you won't have to find additional feed in a drought. However, it is likely that even with the best of management you will sometimes need additional feed. The typical strategy during a drought is to purchase feed, like hay or grain. But since the price of feed usually skyrockets during a drought, this can be a short trip to bankruptcy. This chapter deals primarily with often-overlooked forage sources that can often be accessed for little or no money.

material left in a field after a silage cutter has gone through certainly gives the impression that you "got it all," that is only part of the story. To make good silage it is necessary for a minimum amount of lactic acid to be formed as a preservative. The bare minimum of the gross energy in a feedstuff that must be sacrificed through fermentation to produce enough lactic acid to preserve silage is a whopping 20 percent. In other words, if you harvest 10 tons per acre of forage, 2 tons per acre of it will be lost and converted into heat, water, and carbon dioxide.

Silage harvesting is quite expensive and intensive in both machinery and labor. Additionally, silage is about two-thirds moisture, and thus heavy to haul around. Once the seal on a silage pile is broken, feeding must continue until the pile is consumed, as decay rapidly sets in after exposure to air. Silage harvest puts all the plant material in the pile, including low-quality stem material that may have very little feed value. It costs the same to harvest a pound of stalk as it does a pound of grain, though the amount of animal performance that can be achieved by feeding each can be vastly different. Silage harvest also leaves the soil quite bare and exposed to erosion, sunbaking, and evaporation if it is not immediately planted to a cover crop.

GRAZING PROS AND CONS

Grazing is often the least utilized method of harvesting a failed crop for several reasons. First, there is an assumption that grazing is an inefficient method of harvest. This is usually true, but if you use modern fencing tools and grazing techniques, grazing can actually be more efficient than mechanical harvest. Second, if animals are given free access to a starchy grain crop like corn, they will selectively eat a nearly 100-percent grain diet during the first part of the grazing period. This will almost certainly cause rumen acidosis, which dramatically reduces the ability of the rumen to digest forage, causes liver damage, and can even be lethal. An acid pH kills fiber-digesting microbes in the rumen.

Strip grazing with moveable polywire can be used to force animals to consume all of the material in a paddock rather than select the highest-quality items only. This can be particularly useful in a grain-crop residue pasture, when you want to moderate the intake of spilled grain to prevent acidosis. Daily strip grazing results in a balanced diet of grain and forage material.

Using strip grazing to ration out the forage solves both problems. Strip grazing, in which you move the fence to give access to only one day's worth of feed at a time, ensures that animals have to eat a balance of plant parts in addition to grain, so they cannot overeat grain. With limited grain consumption, the ability of the rumen to digest forage is never compromised, and animal performance is increased. (This is similar to spaced bale grazing, described in chapter 2, but there you use strip grazing to ration out bales; in this case you are using it to ration out standing crops.)

Strip grazing also ensures that a small amount of grain is available each day during the entire grazing period, not just the first part. In addition, it also reduces forage losses from trampling and manure fouling.

While it makes complete sense to use failed grain crops as emergency forage, one big reason it is so seldom done is that crop insurance usually forbids the practice. The time to find out if grazing a failed crop is allowed is *prior to signing the contract*, not when you're scrambling for feed. Communicate with your insurance agent *before* the drought.

Acidosis

There are some cautions to grazing failed crops. Excess starch intake can be a hazard on corn and sorghum residue, as well as any other starchy grain crop. When too much starch is eaten at one time, the rumen can become very acidic, which can wreak havoc inside the animal. It is important to prevent this. The traditional method for preventing acidosis is to gradually adjust animals to a high-starch diet, allowing them a small amount of starch on the first day, then increasing by about a pound of grain a day for a month,

allowing the rumen microbial population to adapt. This is easy to do in a feedlot, where you control the ration. It is not so easy in a grazing situation. The answer is to strip graze to limit the per-day intake of grain. This is a very sound strategy to prevent acidosis.

In addition to using strip grazing to manage intake, there are some feed additives that can prevent rumen from becoming too acidic. Sodium bicarbonate or calcium carbonate can counteract acidosis, as can ionophores such as monensin (Rumensin trademark) and lasolocid (Bovatec trademark). Biochar can also absorb excess acid. There is also a product called Lactipro (available from MSBiotec; see Resources) containing live microbes that can be used to more rapidly adjust the rumen to a high-grain diet. This is given as a drench and contains the organism *Megasphaera elsdenii*, which helps keep the rumen from becoming too acidic when a high amount of starch is eaten all at once.

Nitrate Toxicity

Another risk can be nitrate toxicity. Grain crops in drought conditions often fail to metabolize the nitrate they take up, so it remains in the plants. Small amounts of nitrate are beneficial in a ruminant diet (as well as human diets, we are now finding; part of the reason kale is so nutritious is largely due to its high natural content of nitrates), but large amounts can be toxic. Excess nitrate is converted into nitrite in the digestive system, and when absorbed into the bloodstream can bind to hemoglobin in the blood and render it incapable of carrying oxygen. The result is slow asphyxiation. The telltale postmortem symptom of nitrate toxicity is brown-colored blood, due to having no oxygen content.

It is easier to prevent nitrate accumulation than to manage around it after the fact. One of the primary reasons plants end up with excess nitrate is that they lack some essential mineral nutrient needed to convert it into protein. In my experience, the nutrient most often lacking is sulfur. Plants need sulfur in an amount roughly equal to 10 percent of the amount of nitrogen applied to a crop. It used to be that most of the United States received sulfur in the form of acid rain, resulting from the combustion of coal and diesel fuel. Due to the Clean Air Act, however, coal power plant exhaust is now "scrubbed" to remove sulfur, and diesel fuel has sulfur taken out of it. As a result, sulfur deficiency is now very widespread.

Phosphorus is also necessary to convert nitrate into protein, and in higher pH soils zinc and iron are often lacking for this conversion. Managing soil so that drought stress is minimized also reduces nitrate accumulation, so following all the advice in chapters 2, 3, and 4 of this book can also reduce your risk of nitrate toxicity.

Nitrate can be minimized in the following manner:

Test for nitrate prior to feeding. Most veterinarians have a colorimetric "quick test" that can determine the presence and approximate concentration of nitrate in plants. It consists of an eyedropper full of reagent that, when dropped on the cut stem of a plant, turns a bright color in the presence of free nitrate; the darker the color, the higher the nitrate. For more accurate determinations, every feed testing lab can conduct nitrate testing.

Test water sources along with feed. It is particularly important to test well water, which can be high in nitrate, especially in drought years.

Minimize consumption of lower stalks. The vast majority of the nitrate in a plant is located in the lower part of the stem. Usually, two-thirds of the total nitrate in a plant is located in the bottom 6 inches of the stalk. Any harvest method that results in this part of the stalk being consumed increases the likelihood of nitrate toxicity. Grazing animals usually reject this part of the plant unless there is severe overgrazing. If plants are baled and ground up, an animal is more likely to eat a large part of the lower stalk.

Dilute high-nitrate feed. Dilution is the solution to pollution. Any low-nitrate feed fed as a supplement dilutes the impact of a high-nitrate feed. Feeds usually low in nitrate include mature grass hay (especially unfertilized native grass hay); legume hay; grain; and grain byproducts, like distiller's grain, soyhulls, and wheat middlings.

Adjust animal diets — gradually — to high-nitrate feed to allow rumen microbes that can metabolize nitrate to increase over time.

Use a bolus called Bova-Pro. This contains bacteria capable of metabolizing nitrate into protein in the rumen. This product was developed by researchers at Oklahoma State University (see Resources). Their research shows it can reduce the amount of nitrite reaching the bloodstream from a high-nitrate feed by about 40 percent.

Convert high-nitrate feed into silage to reduce the nitrate potential by as much as two-thirds.

Prussic Acid

Sorghum grazing requires additional precaution against prussic acid. Found in young sorghum shoots, this compound can break down into cyanide when the cells are crushed, as when chewed, and particularly when the plants are frozen. Prussic acid is concentrated in young regrowth; regrowth more than 18 inches tall seldom has any toxicity issues. Once a frost has occurred, the cells are broken and the plants are highly toxic until the tissue has dried out. Cyanide gas escapes as the plant dries. Dried plants are almost always safe to graze; mechanically harvested plants are safe after a month or so of storage.

One way to avoid grazing small regrowth of sorghum, prior to a killing freeze, is to rotationally graze as follows:

1. Subdivide a field into four or more paddocks.

2. Turn animals into the first paddock after the plants are at least 24 inches tall.

3. Graze until the leaf material is gone but the stems remain (reducing nitrate risk by leaving the stems ungrazed).

4. Move to the next paddock and repeat.

5. Return to a paddock only when regrowth exceeds 18 inches in height.

CROP RESIDUE

Grazing crop residue is a common practice in many areas, but even where it is commonplace only a small fraction of the grazing potential is realized. This is primarily due to poor grazing management.

As discussed earlier, when animals are given access to an entire field of crop residue they will trample most of the feed looking for dropped grain. Not only does this lead to large trampling losses, it also results in a diet of nearly pure grain during the initial part of the grazing period. The rumen then becomes too acidic, which kills off the microbes necessary for the digestion of cellulose. While the grain is digested well, the forage is not, and animal performance is far less than it should be, particularly at the tail end of the grazing period.

The solution to this is, again, strip grazing, with a fence moved on a daily basis. Beginning at the water source, simply use two portable fence reels and step-in posts to move a day's worth of feed away from the water. For someone new to the practice, it can be difficult to estimate how much area to allow the animals to access each day. Place some low-quality hay in a feeder near the water source to provide a safety valve in case the area is not sufficient. Using low-quality hay, rather than good hay, ensures that the animals eat the crop residue in preference to the hay, and they eat the hay only when the crop residue is thoroughly consumed. It is normal for animals to eat a small amount of hay no matter the feed quality, but if hay intake gets high, it is likely because the grazing area is too small for the grazing period.

It is important to move the fence daily where grain is available. Grazing periods longer than 1 day per paddock allow grain consumption to be excessively high on some days, causing acidosis. Once acidosis occurs, it may take as long as 6 weeks for the rumen to return to normal.

The value of strip grazing has been proven in both practical experience and in research by every agriculture university in the Corn Belt. The results of strip grazing are remarkably uniform and clear: the grazing days per acre are doubled, and cows come off in better body condition than their sisters that are allowed the entire field at once.

My Strip-Grazing Pitch

When I pitch the idea of strip grazing to farmers and ranchers, you would think I asked them to donate a kidney. "Oh my gawd! I don't have time to move a fence every single day. You have no idea how busy I am! I can't possibly fit that into my day."

Then I ask what they are so busy doing. "Well, I get up at 5 a.m. to start the tractor and warm it up, then I have to get hay bales, and grind them, then I have to scoop silage, then put it all into the feed wagon, then haul all that to the cows. And every day something breaks down and I have to use a cutting torch and welder to fix it, then get to the parts store before they close at five. By the time I do all that, the winter daylight is gone."

When I tell them that moving a fence daily *replaces* all that work, they look at me like I have a third eyeball coming out of my forehead. Apparently, it just seems too good to be true. I then ask them how many acres they farm, and offer this proposal: "If the landowner right across the road offered to rent you that many additional acres of land, would you say yes?" Never will a farmer turn down access to more land.

Then I ask, "What if he offered it to you rent-free? And what if he said he'd pay for all cash inputs, and you just had to provide labor?" Their eyeballs bulge out and they almost drool when they tell me, "Yes, yes, yes, I would say yes!"

I reply, "*No*, you just told me you would have to say no, because you are too busy. That additional acreage would double your workload. You don't have time." And their answer: "But, but, but, I could double my production with no additional costs, just by working harder."

That is exactly how strip grazing works on crop residue, except your productivity doubles with about 10 additional minutes of labor each day. For the life of me I cannot understand why this is not a universal practice wherever crop residue is grazed.

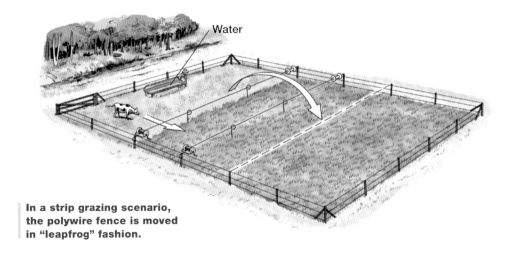

In a strip grazing scenario, the polywire fence is moved in "leapfrog" fashion.

Water

CROP RESIDUE AND NUTRITION

Another key to improving utilization of crop residue is to understand its nutritional limitations. Crop residue is typically very low in protein. Corn residue is often much lower than the minimum of 7 percent protein needed for good rumen function.

Note: I often use the terms *crop residue* and *corn stalks* interchangeably, because the most-available crop residue in the United States is from corn. The rules vary for other crop residues, but most are typically low in protein. Supplementing even a small amount of protein can dramatically improve animal performance. Intake increases with protein supplementation, and even though animal performance is dramatically increased, the carrying capacity of a stalk field can in fact be slightly less when protein is supplemented.

Ammoniation

One method of adding protein to a ruminant diet with harvested crop residue is to put bales under a sealed plastic tarp and inject anhydrous ammonia in the amount of 3 percent of the weight of hay. This process, called **ammoniation**, breaks the bonds between indigestible lignin and digestible cellulose, increasing the digestibility of crop residue by roughly 10 to 20 percent. In addition, rumen microbes can utilize ammonia to make protein, which can then be utilized by the animal. In feeding trials, ammoniated wheat straw gives similar performance to fair-quality grass hay, and the results with ammoniated corn stalks are slightly better.

If the manure thus produced is deposited where it can be put to good use, the anhydrous ammonia does double duty, first as a feedstuff, then as a fertilizer. A pound of anhydrous ammonia run through a rumen is roughly equivalent to 2.8 pounds of protein, or about the same amount of protein found in 6 pounds of soybean meal. This is in addition to its value in improving digestibility of the forage it is applied to. Since a pound of ammonia usually costs about $0.50, and 6 pounds of soybean meal is

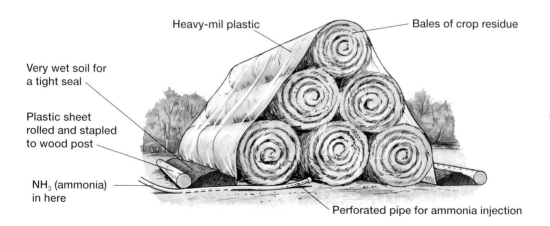

Heavy-mil plastic

Bales of crop residue

Very wet soil for a tight seal

Plastic sheet rolled and stapled to wood post

NH_3 (ammonia) in here

Perforated pipe for ammonia injection

Ammoniation of low-quality crop residues can greatly enhance their digestibility by ruminant livestock. The process involves placing an airtight plastic cover over the bales (the end is open in this diagram for clarity, but in actual practice the entire stack is covered) and slowly trickling anhydrous ammonia into the stack over several hours.

typically three times that amount, ammoniation is a very cheap way of providing protein to a ruminant.

Ammonia is not a complete feed, obviously, nor even a complete protein supplement. It has no energy content, no vitamins, and no minerals. These have to come from the rest of the ration. Ammoniation works best when the feed is supplemented with a small amount of distiller's grain (with solubles) and/or legume hay. These contain branched chain amino acids that rumen microbes have difficulty synthesizing from ammonia. Some supplemental sulfur (which distiller's grain has in abundance) is also necessary for optimum performance.

Ammoniated crop residue works best when fed as only a portion of the total ration. I have friends who use ammoniated wheat straw as a supplement to crop residue pasture, because the palatability is less than that of leaves and husks (the good parts of crop residue) but more palatable than the stalks themselves. Once consumption of ammoniated straw begins in earnest, it is time to move.

The fermentation extract of the fungus *Aspergillus oryzae* (trade named Amaferm, and marketed by Vitaferm and other companies) contains enzymes that can break down plant fiber in a manner that rumen microbes themselves cannot. In soil, fungi are the natural decomposers of lignin, because they also contain enzymes, but the rumen of an animal has relatively very few fungi. Putting these enzymes in the rumen can dramatically increase utilization of poor-quality roughage. Research indicates a big boost in cellulolytic activity in the rumen. This is a product I have used extensively on my own ranch, and I never cease to be amazed at how effective it is.

How Straw Treatment Affects Weight of Cows

As can be seen from the research below, ammoniated straw can be utilized as an adequate maintenance feed for nonlactating cows, although better results are achieved when it is only a part of the total ration.

TREATMENT	STRAW INTAKE (LB)	COW AVERAGE DAILY WEIGHT CHANGE (LB)
Wheat straw +7 lb alfalfa	14.8	-0.27
Ammoniated straw	26.1	+0.10
Ammoniated straw +7 lb alfalfa	19.7	+0.40

Table 7.1

Good cattle ranchers can estimate the digestibility of their forage by observing the "patty index," or the thickness of the cowpie. The thicker the cowpie stacks up the more undigested fiber it contains. The first time I used this product I had cattle on sorghum stalks, and their cowpies were probably 6 inches thick the day before rolling out a Vitaferm tub. The following day, the cowpies looked as though the animals were on green spring grass, with the cowpies maybe an inch high at the most. I was thoroughly impressed, and I now use the product routinely any time my animals are utilizing poor-quality feed. The research I have seen on this product indicates it is very effective at boosting digestibility of low-quality forage but offers much less value when used with higher-quality forage.

Methane and Ionophores

Ionophores are feed additives that cause a shift in the rumen microbial population away from methane producers and toward more energy-efficient microbes. Methane contains a lot of energy: it is the primary compound in natural gas. When ruminants belch methane there is a great deal of energy that escapes the rumen; thus the microbes that produce methane result in less animal performance per unit of feed energy than microbes that do not produce methane. There are methane-eating microbes (**methanotrophs**) in the rumen, but since methane is a highly volatile gas, most of it escapes out of the rumen before it can be utilized. Practices to prevent methane production improve feed efficiency and animal performance as well as reduce production of a potent greenhouse gas.

The two most common ionophores are monensin (Rumensin, trademark of Elanco) and lasalocid (tradename Bovatec). A summary of 24 trials (see References) reported that monensin fed to pastured cattle reduced feed intake by 3.1 percent and increased the average daily gain by 0.09 lb per day, for a feed efficiency improvement of 15.3 percent. This means 15.3% less hay that needs to be fed or 15.3% less pasture consumed to get the same results as without the ionophore.

Methane and Biochar

Since methane is an energy source, and there are methanotrophs (microbes that eat methane) in the rumen that are in turn digested by the animal, a way to trap the methane in the rumen rather than let it be belched out could also potentially improve feed efficiency and simultaneously reduce the emission of a potent greenhouse gas. Feeding animals a small dose of biochar daily can adsorb

methane and keep it in the rumen fluid for access to microbial activity. This is further discussed in chapter 8.

WINTER COVER CROPS

Interseeding winter annual cover crops into standing corn crops can dramatically improve the value of crop residue in a couple of ways. First, winter annual cover crops add to the total biomass of feed available by growing after the cash crop has died. Second, the winter annual cover crops are in a high-protein vegetative state during the typical stalk-grazing period, and thus help solve the protein-deficiency problem.

To use this blend effectively it is critical to utilize strip grazing or rotational grazing.

| Interseeded cover crops in corn

If animals are allowed free access to an entire corn stalk field that has had a cover crop of rye seeded into it, you'll observe this sequence of events:

1. First, the animals will trample half the field foraging on dropped ears of corn. This will usually trigger some level of acidosis.

2. Then, they will trample the field again, picking at whatever green material is available. This will be cleaned up in very short order, as the cover crop biomass is generally fairly light in comparison to the amount of crop residue available.

3. Finally, the cows eat the husks, then the leaves. By this point they are short on protein, and performance is less than optimal.

Strip grazing or rotational grazing this crop blend results in a daily consumption of some grain, some high-protein green cover crop, and a lot of crop residue; in other words, a perfectly balanced diet for a cow.

Let's say you have a corn field with harvest loss of 5 bushels an acre, or about 280 pounds of grain per acre. Let's round that up to 300 pounds to make the math easy. Let's also assume we have 1,500 pounds of green cover crop available, and 3,000 pounds an acre of husks and leaves. Now let's put a cow on this acre and assume she eats 30 pounds a day. If we let her roam free over the entire field, her diet will be nothing but corn for 10 days, green material for 50 days, and husks and leaves for 60 days. She gets acidosis in the first 10 days, and it takes 6 weeks for her rumen to heal. By this time, the only thing left is the low-protein crop residue she didn't trample earlier.

Alternatively, we can limit her grazing area by moving a polywire daily, and she will get about 2 pounds of grain (not enough to get acidosis), about 9.4 pounds of high-protein green cover crop, and 18.75 pounds of husk and leaf daily. This amounts to a nearly perfect brood cow ration, and her rumen stays healthy and ready for spring grass when it occurs.

COVER CROPS AND DROUGHT

Cover crops can be extremely valuable forage during drought, although (as it is often pointed out to me) cover crops typically don't grow any better during a drought than cash crops do. The hardest cover crop to grow usually is the very first one on a piece of ground; once a mulch is established, each successive cover crop (and cash crop) gets easier and easier to grow. A nice surface layer of well-decayed organic material protected by a thin layer of fresh residue makes for a very forgiving seedbed that can sprout newly planted seed even in very dry conditions.

Unfortunately, many people try planting cover crops only after the drought has already progressed, and the results are disappointing. These people then say, "Cover crops just don't work around here," based on their experiences from the last drought. It is funny that they don't say "Corn doesn't work around here" or "Soybeans doesn't work around here," based on the fact that they failed that year as well. Ironically, it was during the 2012 drought that the positive results from cover cropping became most apparent to me. Crops planted into land that had been cover cropped for several previous years actually fared fairly well that year.

I can thus give two pieces of advice about cover crops:

Don't wait until the next drought to try them. Start now and build that desirable organic layer during the good years so you can benefit when the drought does occur.

Don't limit yourself to your own acreage when planting them. There are many farmers who would like the benefit of cover crops on their land but have no livestock to utilize the cover crop for grazing and thus cannot make the cover crop pay for itself that way. Often the cheapest grazing you can get is from cover crops planted on neighbors' land for the cost of the seed, your time to plant them, and a pittance of a land rent during the offseason.

There has been enough positive press about cover crops that many farmers would like to try them, but those without livestock usually do not because they are less likely to see a short-term cash return without grazing. These are precisely the neighbors you want to partner with in your drought preparation. Many farmers who have no livestock will leap at a chance for someone to pay for a cover crop on their ground in exchange for grazing rights. There are others, however, who have concerns about moisture use by a cover crop, and by compaction effects from grazing on their cropland.

COMMON CONCERNS ABOUT COVER CROPS

Regarding moisture use, there is no question that cover crops use more moisture than untilled fallow ground, to the tune of about 1.5 to 2 inches more water per season. However, once the cover crop has been terminated, that ground immediately becomes more moisture-efficient than the ground that was not cover cropped; the mulch layer slows evaporation and improves infiltration, and the decaying organic matter improves the water-holding capacity of the soil (if only slightly). Research in my home state of Kansas indicates a cover crop in wheat stubble uses about 2 inches more moisture than bare wheat stubble; however, by the next July (when it really counts) the ground with cover crops will, in most years, have more moisture than the uncovered ground, and this will remain so going forward.

Regarding compaction, hundreds of studies have been done by university researchers that indicate there is minimal, if any, yield loss from grazing cropland during the winter. One of the best summaries I have seen on this matter was done by the University of Nebraska (see References). And yet, some people will not be convinced by any amount of research. If this is what you face when negotiating land use, you can at least partially alleviate their concerns by planning ahead to avoid grazing compaction during wet weather.

Plan ahead by setting up a "sacrifice area" where animals can be pulled off when the soil gets too soft to support livestock hooves. Find a small grassed area or adjacent pasture, or just an area of the field that is separately fenced where animals can be confined

What Cover Crops to Plant, and When

Whenever there is idle land, that land can be planted to a cover crop, assuming the growing window is long enough to recoup the expense of planting. Here are some options to produce forage during common planting windows, arranged from highest yield to lowest:

Planted early spring (for grazing or harvest 60–90 days later)

- Oats, spring barley, spring triticale

- Spring field peas, chickling vetch, lentils

- Rapeseed, turnips

Planted early summer (for grazing or harvest 45–60 days later)

- Sudangrass or sorghums

- Brown midrib corn

- Millets (pearl millet is best grazing; Japanese, browntop, and foxtail millet offer lesser value; and proso millet has virtually no forage value)

- Crabgrass

- Teff

- Sunn hemp (any variety for sheep, goats, or deer; Tropic Sun variety for cattle)

- Okra

- Cowpeas

- Lablab

- Forage soybeans

- Guar

- Korean or striate lespedeza

Planted late summer/early fall

All the early summer options still work in this window but with greatly diminished yield potential compared to earlier planting. Still, summer annuals may outyield cool-season annuals even when planted late. Summer annuals should have 45 to 60 days growing prior to average first frost.

and fed hay until conditions improve. This can be an ideal location to use bale grazing or spaced bale feeding, to the great benefit of the soil in that location over the long term.

Actually, two of the best ways to remedy compaction are applying manure and growing a cover crop, and one of the ways to make compaction worse is by running heavy equipment (combine and grain cart) on a field. But for some reason, people think livestock cause compaction while combines magically do not, even though a new John Deere combine

with a 350-bushel tank full of grain can weigh 53,000 pounds on the front axle alone. Since the tires on that axle cover only a few square feet, that is like stacking 40 normal-size cows on top of each other on every tire print left by the combine.

Cool-season annuals listed below are for planting up to 30 days prior to first frost.

- Turnips, forage radishes, kale, rapeseed, and hybrid brassicas (fall grazing only, begin 60 days after planting, winterkill at 16°F/–9°C)

- Oats, spring barley, spring triticale (fall grazing only, begin 60 days after planting, winterkill at 16°F/–9°C)

- Spring peas, chickling vetch (fall grazing only, winterkill at 18°F/–9°C)

- Winter barley, winter rye, winter triticale, winter wheat, annual ryegrass (these plants typically overwinter to provide spring grazing as well)

- Winter peas, hairy vetch*, crimson clover, balansa clover (these plants typically overwinter in USDA winter hardiness zones 5 and above to provide spring grazing as well, though they are not as winter hardy as the winter annual grasses)

Planted late fall, around time of first frost (limited fall grazing, good spring grazing)

- Rye, triticale, annual ryegrass, winter wheat, hairy vetch*

Planted very late fall, after hard freeze (spring grazing only)

- Rye

*Hairy vetch has alkaloids to which certain breeds of cattle (primarily black Angus and Holstein) and dark-hided horses can be fatally allergic. This allergy seems to occur within about 5 percent of those breeds, and at a much lower rate in other breeds. It is a gruesome death when it occurs, and there is no remedy. For this reason I am much less "gung-ho" on grazing hairy vetch than I used to be, though I have grazed it for years with no incidence, as have many people I know. Many other forages have toxins that we need to manage, but hairy vetch toxicity is poorly understood at this time and therefore difficult to manage.

BLEND SPECIES TO OVERYIELD

When farmers first try cover cropping, they usually try to plant the "best" cover crop species for their situation. Most experienced cover croppers are now using a blend of several species together, from several different plant families. This follows the concept of **overyielding**, when a blend yields more than the weighted average of all the species in the blend.

For example, let's take a blend of 50 percent sorghum and 50 percent cowpeas, compared to a full stand of each. Let's assume a pure stand of sorghum yields 6 tons an acre, and a pure stand of cowpeas yields 2 tons an acre. The predicted yield of a 50/50 blend would be 50 percent of 6 tons of sorghum (or 3 tons), plus 50 percent of 2 tons (or 1 ton) of cowpeas, for a combined total yield of 4 tons.

In reality, the yield of the blend usually far exceeds this predicted yield and comes close to the yield of the most productive species in the blend. This is due to several factors:

- Root layering, in which multiple root types and depths are more efficient at extracting water and nutrients than a single root type.

- Leaf layering, in which different leaf types and heights capture more total sunlight than a single leaf type.

- No single insect or disease or nematode is likely to afflict all the species in a complex blend (except perhaps grasshoppers), and therefore more yield is achieved.

Finally, animal performance is typically better when there is dietary diversity.

TREES AS FORAGE

At one time in Kansas, there was an attempt to make a silk industry. The food of the silkworm, the mulberry tree, does exceptionally well in Kansas, where there are even native species of the tree. The industry collapsed before any commercial silk was produced, as it became clear that — even over a hundred years ago — Americans would not work as cheaply as Chinese workers would. The silk industry disappeared, but the mulberries thrived and spread.

When my grandparents married in 1945 and were looking for a farm to buy, they found a place that was cheap because it was covered in mulberry trees. It took a lot of work to convert those mulberry acres to other uses. During the drought of the 1950s, however, those trees were the only green vegetation on the farm. Every day, my grandfather went out with his chain saw and felled one tree for every cow, and the cows greedily ate every leaf on them. This was an emergency measure, but the cows in fact did better than they ever had before. It turns out that mulberry leaves are quite nutritious, with protein similar to alfalfa's and energy content similar to that of corn silage. In other words, they are dairy-quality feed. Those mulberries saved my grandparents' farming operation.

When I was first ranching and faced a drought situation myself, I felled some trees in a wooded pasture in order to supplement the cows' diets. I found out they would eat a number of trees, not just mulberries. They ate elms, cottonwoods, willows, and even the thorny Osage orange and honey locusts (when felled, since only the lower branches tend to have thorns). There seem to be only a few species they do not eat, especially

Cattle find tree leaves to be both palatable and nutritious, as with this mulberry tree felled to provide access to the leaves for emergency feed.

when they are trained to recognize tree leaves as food.

The biggest problem is keeping the cows away from the trees while you are dropping them, to avoid injury. In those days, I was not using paddock subdivisions, but now I could simply fell trees in a paddock before letting cattle in there.

Obviously, cutting trees for emergency forage cannot be an every-year practice, unless you have a large enough grove. But if cutting trees for pasture improvement is a standard practice in your area, I would advise doing it during a drought period when the leaves can be put to good use, rather than in the winter, as is typically done.

Use Trees for Drought Protection

Trees can also become a planned drought reserve. You can easily establish cottonwoods for shade by inserting freshly cut poles into the ground. These will quickly grow into shade trees and when needed can be converted into feed with a chainsaw.

Willows can be used to tremendous advantage, offering exceptional protein content and palatability. They also host nitrogen-fixing bacteria, not on their roots but in their leaves and stems. This allows them to thrive in the low-nitrogen soils created by standing water — a rather unique ability.

If you choose to put into practice some of the landforming ideas we have discussed, such as creating swales to impound water on the contour, an ideal vegetation to plant in them would be something that can tolerate standing water and has exceptional forage quality in late summer, when drought typically occurs. This describes willows perfectly.

Weeping willows have drooping branches that are easy for cattle to reach, for the most part, and in extreme forage shortage the top branches can be felled for easier access.

WETLANDS AND PONDS

As discussed earlier, ponds and wetlands should be fenced to prevent access by livestock, because unrestricted access can quickly ruin their value not only for domestic animals but also for wildlife and ecosystem function. Nevertheless, many wetland plants can provide good forage in an emergency. You can reserve these areas for emergency use only by fencing them off but providing an access gate. See more about ponds in chapter 5.

WEEDS

As I was driving through the epicenter of the 2012 drought in southwest Kansas in late August, I saw something that struck me as odd. On a quarter section of land there was a pasture next to a field of wheat stubble. In the pasture (which was so overgrazed that it looked like bare dirt) a tractor was hauling bales of hay to a bunch of skinny cows. In the wheat stubble I saw a thick growth of kochia, pigweeds, and volunteer wheat. It would have made complete sense to me to simply put an electric fence around the wheat stubble and open a gate. That kochia/pigweed/volunteer wheat probably had three times the protein content and 50 percent more energy content than the hay the farmer was feeding, and it was free for the taking. But no, as one tractor hauled the hay to the pasture, another was pulling a sprayer across the weedy field. Thousands of dollars' worth of nutritious feed was being deliberately destroyed and replaced with hay worth thousands of dollars itself, purchased at a historically high price.

Many weeds are surprisingly nutritious and remarkably palatable. There are some weeds that are palatable to cattle, others to sheep, others to goats, and still others to horses, but nearly every weed is eaten by some domestic animal. Cattle do not like yellow foxtail, for example, but horses love it. Cattle eat pigweeds when they are young and in fact prefer them to almost any other forage. When I turn cattle into a new pasture paddock, they eat dandelions before they touch any of the species I actually planted. If dandelions produced any tonnage we would be planting them for pasture instead of fighting them. I am not suggesting that we should plant weeds for forage, but we might as well take advantage of them when an opportunity or need arises.

Some weeds that cattle find highly palatable: Pigweeds, kochia, giant foxtail, green foxtail, crabgrass, barnyard grass, fall panicum.

Some weeds that cattle will eat when managed with daily moves, but not otherwise: Sunflowers, giant ragweed, velvetleaf, lambsquarters, yellow foxtail, marestail.

Some weeds that cattle never seem to eat: Cocklebur, thistles, buffalobur (*Solanum rostratum*).

Note that almost all weeds are much more palatable when young than when mature and woody. Sheep, goats, and other livestock find many weeds palatable that cattle do not.

DORMANT GRASS

I live in an area where native warm-season grasses, like big bluestem, switchgrass, Indiangrass, and sideoats grama, are the major pasture resource. These pastures are typically used only for summer grazing and are almost always idle during the winter. The protein content of these grasses in the winter is too low to support good animal performance, since rumen microbes need about 7 percent protein to reproduce, and these grasses average about 3 percent in winter. Nevertheless, these same grasses fed millions of bison for thousands of winters prior to our arrival on this continent. So yes, they can be used for winter grazing, and yet millions of acres are left completely unutilized during the winter every year. These acres are opportunity, if managed correctly.

SETTING THE ANIMAL UP TO UTILIZE DORMANT PASTURE

The first hurdle to taking advantage of winter pasture is having an animal whose nutritional needs are minimal enough that the low protein content is not a problem: in other words, a mature, non-lactating animal. Dormant native grass is not milking-cow feed, even if supplemented. The reason bison could use this grass to advantage is because they calved in May, after the new grass had greened up. In winter, they had weaned their calves and therefore were at their lowest needs of the year. Most ranchers today have shifted their birthing season to midwinter, shifting the highest nutrient-need period to a time when it is most difficult to supply — and basically impossible to supply with dormant warm-season grass.

The biggest step in making dormant grass an asset is to shift the birthing season to the beginning of the natural high-quality grazing season. This was the single most positive move I ever made on my own ranching operation. Table 7.2 on the facing page illustrates the needs of a beef animal by production stage.

SUPPLEMENTING PROTEIN

The second hurdle is to overcome the protein deficit. This is easy enough with the proper protein supplementation. The ideal protein supplement for dormant warm-season grass is high in protein with little, if any, starch. Starch fermentation reduces fiber digestibility, as illustrated in Table 7.3.

Note the poor fiber digestibility and reduced forage intake with the low-protein but high-energy 12 percent crude protein supplement (see Table 7.3). This difference is reflected in animal performance as well, as illustrated in Table 7.4.

What Supplements to Provide and How to Do It

Supplements that fit this high-protein / low-starch category include legume hay and distiller's grain. Oilseed meals, such as soybean meal or cottonseed meal, are similar but not quite as good. The amount of supplemental protein needed per day is roughly a pound for a mature dry cow. This translates to a daily supplement of about 6 pounds of 16-percent protein legume hay, or 4 pounds of 25-percent distiller's grain, or 2 pounds of 48-percent soybean.

Table 7.5 illustrates the slight superiority of a high-protein forage supplement (alfalfa cubes) over a cottonseed meal (CSM) and barley cube when fed at equal levels of protein and energy with dormant winter tall-grass range.

There are two ways to provide this supplementation: self-feed and hand-feed. Self-feeding involves using an intake limiter, usually salt, to ensure that animals eat only enough supplement to meet their needs and no more. Cattle typically eat a maximum of a pound of salt daily, so a protein supplement in meal form can be blended with salt and fed free choice with only limited consumption. (Guidelines for making a salt-limited supplement are covered in an excellent extension publication by the University of Nebraska; see Resources.)

As a starting point, a meal with 20 percent salt and 80 percent distiller's grain will result in adequate protein content, as will a meal with 33 percent salt and 67 percent soybean meal. These levels may need adjustment to achieve the desired intake. Also, it is critical that animals on a salt-limited ration have full access to clean water that is low in salt content. This is especially important in winter, when freezing of water supply systems can be a problem.

The other way to deliver limited protein is by hand-feeding. One issue with feeding a limited amount is that the more aggressive animals tend to get more than their share, while smaller and less aggressive animals don't get enough. One way to help overcome this is to feed a triple amount of supplement every third day, rather than daily. For example, instead of feeding each cow 6 pounds of alfalfa hay per day, you would feed 18 pounds every third day. The additional protein remains in the animal's system for at least three days after feeding. Contrary to what you might expect, animal performance is actually improved by this process (see Table 7.6). When larger amounts are fed infrequently,

Daily Protein Needs of a 1,300-Pound Cow by Stage of Production

STAGE	DAILY CRUDE PROTEIN NEEDS (LB/DAY)
From weaning until 60 days prior to calving	1.5
60-day period prior to calving	1.8
From calving to rebreeding	2.3
From breeding until weaning	2.0

Table 7.2

How Crude Protein Supplements Improve Digestibility of Dormant Forage

	NO SUPPLEMENT	12% CP	27% CP	39% CP
Forage intake as % of body weight	0.9	0.8	1.4	1.2
Total intake as % of body weight	0.9	1.0	1.9	1.6
Fiber digestibility as %	37.9	29.9	39.9	38.6

Table 7.3

How Supplemental Protein Improves Animal Performance on Dormant Pasture

	13% CP	26% CP	39% CP
Weight loss from Nov. 15 through calving (lb)	192	123	97
Calf birth weight (lb)	74	78	82
Conception rate (%)	86.7	93.3	93.3

Table 7.4

How Winter-Range Supplements Compare

	NO SUPPLE-MENT	CSM/ BARLEY	ALFALFA CUBES
Weight change (lb)	−24.2	+30.8	+52.8

Table 7.5

Note that in the studies shown in Tables 7.3–7.5, all the supplements provided equal energy in the form of starchy grain. The supplemental protein made the difference in the results.

each animal is more likely to get its share, and it will spend less time waiting at the gate for the feed truck and more time grazing.

Yet Another Opportunity for Strip Grazing

I may be sounding like a broken record, but another management method to improve utilization of dormant pasture is strip grazing. Strip grazing forces the animals to eat a controllable percentage of each daily allotment, as opposed to walking the entire area and eating all the best feed during the first few days. This is especially important and useful when incorporating the next idea for improving utility of dormant pasture: interseeding with annual plants.

Intercropping with Annuals

Intercropping pasture involves seeding a cool-season annual forage into a warm-season grass sod. This is common where bermudagrass is the main pasture species, but not in rangeland. Instead, in many rangeland areas, you'll find a practice called "chasing cheat." Cheatgrass is a term for an annual cool-season bromegrass that invades warm-season pastures. It has a brief period of high palatability and protein in the spring (before native grasses begin growth) and then becomes very unpalatable after it forms seedheads.

If animals are allowed access to the entire grazing area, they will walk the full pasture and eat only cheatgrass, completely

How Feeding Interval Affects Calf Crop and Weaning Weight

This table demonstrates the superiority of thrice-weekly supplementation over daily supplementation. Each group was fed the same total amount of supplement during the study period.

GRAZING INTENSITY AND FEEDING INTERVAL	CALF CROP %	WEANING WEIGHT (LB)
Light grazing, fed daily	78	403
Light grazing, fed 3× a week	84	419
Moderate grazing, fed daily	70	387
Moderate grazing, fed 3× a week	72	387
Heavy grazing, fed daily	65	399
Heavy grazing, fed 3× a week	71	397

Table 7.6

consuming it in just a few days. If they are strip-grazed, however, you can manage them to have enough cheatgrass to meet their protein needs and then fill up on dormant grass for the remainder of their diet. The practice of chasing cheat illustrates that there is value to having a cool-season annual in a warm-season sod, if managed with strip grazing.

My experience has been that cool-season perennials do not coexist well with warm-season perennials and often eventually take over. Cool-season annuals are usually less competitive against warm-season plants and often coexist well. A cool-season annual in the pasture acts as a protein supplement, replacing the need for imported feed. A cool-season grass, however, may need additional nitrogen or sulfur to make this practice work (remember, it takes nitrogen and sulfur to make protein).

Legumes Step Up

In areas where annual ryegrass is interseeded into bermudagrass sod, additional fertilizer is applied to feed the additional grass growth. While cost-effective (usually) in higher rainfall areas, on many rangeland areas this practice would never pay. That is why I believe using legumes for this purpose may be the ultimate solution, since they can make their own nitrogen. The legume not only provides enough protein to balance the diet, it also increases the growth of the grasses around it through nitrogen contribution. Incorporating cool-season plants into a previously warm-season pasture also adds to the days that a pasture is photosynthesizing. This should add biomass (and carbon) both above and below ground.

Summer annual forages growing in a stand of tall fescue

Broadcast News

Walt Davis, a rancher I know, spent several years broadcasting seed of dozens of species of annual legumes into his pastures, taking note of which ones worked in which areas. His wife called it "his annual sacrifice to the clover gods." But this shotgun trial-and-error approach, coupled with Walt's careful and diligent observations, has revealed priceless information and led to a great improvement in pastures. Walt has authored several books, including *How to Not Go Broke Ranching* (see References).

Cool-season clovers in a perennial warm-season grass pasture can add nitrogen as well as extend the growing season and act as a protein supplement for winter grazing.

Finding the right species may be tricky and is often region-specific. The ideal legume should be easy to establish, preferably by broadcasting, since drilling into many pasture soils is not feasible due to rocks and rough topography. Also, this legume should not be competitive with the desired perennials, and it should readily reseed. In many areas I have visited, black medic seems to fit these criteria. Farther south, crimson clover, balansa clover, arrowleaf clover, and some of the burclovers and medics seem to work.

Pasture Cropping

The practice of interseeding cool-season plants into warm-season sods is perhaps epitomized by a practice called **pasture cropping**, pioneered by an Australian named Colin Seis and some friends. This involves the planting of cool-season grain crops, like winter wheat, into warm-season pasture sods. While the goal of pasture cropping is to produce a grain crop and a pasture crop from the same acre in the same year, it has also been shown to cause a rapid increase

in soil carbon. This is because it adds to the number of days photosynthesis is active on a land area. Since we often consider a perennial grass sod to be unbeatable for improving soil carbon, it is noteworthy that this practice has improved upon that system.

The fact that pasture cropping increases carbon storage has significant potential not only for improving pasture productivity over much of the world's rangelands, but also for sequestering a great deal of additional carbon in soils we previously thought were "as good as they are ever going to get." Rangelands cover 70 percent of the world's land surface outside of Antarctica (as compared to the 12 percent that is cropped), so there is obviously great potential for mitigating atmospheric carbon dioxide through management of Earth's rangelands.

LAWN CLIPPINGS

During a severe drought, it's not unusual to take a drive to town and see people watering their lawns. This grass will get mowed off, and much of it will be bagged and hauled to a landfill. What a waste! What if these clippings could be used as a feed resource? Since lawn clippings are almost always mostly leaves, the quality is usually quite high. While the yield from any one lawn would be a drop in a bucket for a livestock operation of any scale, the combined haul from numerous lawns could be substantial.

There are several potential sources for grass clippings, including commercial lawn-mowing services, golf courses, and city park maintenance operations. Since many lawn services pay a tipping fee to dispose of clippings at landfills, you may be able to have this material delivered to you for free, to the delight of both parties. If delivery is not possible, perhaps you can drop off a trailer at the lawn service and pick it up when it's full. If you have a family member or neighbor who works in town, or a teenager driving to school, you may be able to sweet-talk them into daily delivery. Lawn clippings begin to spoil quickly, so it is important to feed them daily.

HAYING AREAS THAT WILL OTHERWISE BE MOWED

Many areas are mowed simply for sake of appearance, such as roadsides and park areas. Often the mowing is considered an undesirable chore and can represent a considerable expense. It may be possible to obtain haying rights by offering the benefit of reducing labor and mowing costs for someone else.

CONTROLLING GRASSHOPPERS

Grasshopper infestations usually worsen in dry years, which makes forage shortages even worse. Grasshoppers can devour a huge amount of vegetation, and controlling them can ensure better feed availability.

One obvious way to control grasshoppers is to have the local co-op spray insecticide. Though effective, this has the drawback of killing all the insects out in your pasture, over 99 percent of which are beneficial or benign.

Public roadsides may offer a source of forage.

It also deprives songbirds and other insect-eating animals of their food source. I used to have a wonderful population of orioles in my pastures until I decided to take out my alfalfa weevil population. Years later, the orioles have yet to forgive me and return. I enjoyed watching all the flying orange acrobats swoop down and eat a good portion of my pest problems, but sadly it no longer happens.

I avoid the broad-spectrum insecticide approach because it seems that removing the natural population of predator insects and spiders almost always ensures that another pest will crop up later. I prefer more targeted approaches. One insecticide that is more specific in its action is called Dimilin. It is an insect growth regulator that prevents molting, so it targets insects with a gradual metamorphosis, and it tends to be safe to insects with a complete metamorphosis, like beetles and bees. It is also quite economical and has fewer off-target effects than most broad-spectrum insecticides. To be effective, it must be applied when the grasshoppers are in the nymph stage.

Another control method that is even more effective is the use of a wheat bran–based bait that contains a dose of the insecticide carbaryl (familiar to gardeners as the insecticide Sevin). This targets insects that feed on the bait but leaves predatory insects alone. It is very effective on both grasshoppers and crickets in all growth stages.

Another method that uses a bran bait involves lacing the bait with the spores of the protozoan grasshopper parasite *Nosema locustae*. This is a very slow-acting control, often taking 3 to 6 weeks to kill. The beauty of this method is twofold: it is highly specific to grasshoppers and crickets (preserving beneficial insects, spiders, and insect-eating birds), and it persists and even spreads, since grasshoppers are cannibalistic and devour their dead comrades, spreading the disease. It is not unusual to get 2 to 3 years of decent grasshopper control with one application. The major drawback of this method is that it *must* be applied early in the grasshopper life cycle to have an effect before the grasshoppers are out of hand.

CHAPTER SUMMARY

There are a number of strategies for feeding animals in a drought; we are limited only by our imaginations. Some of the practices described are as follows:

- Utilize failed crops for pasture.

- Improve utilization of crop residue with strip grazing.

- Plant cover crops for emergency grazing, or to improve value of crop residue.

- Use wetlands for emergency forage when they dry up.

- Fell trees to be used as emergency forage.

- Dormant grass can be used to provide winter forage, if properly supplemented.

- Lawn clippings can be a helpful, though limited supplement to stretch pastures.

- Reduce competition for pasture from grasshoppers.

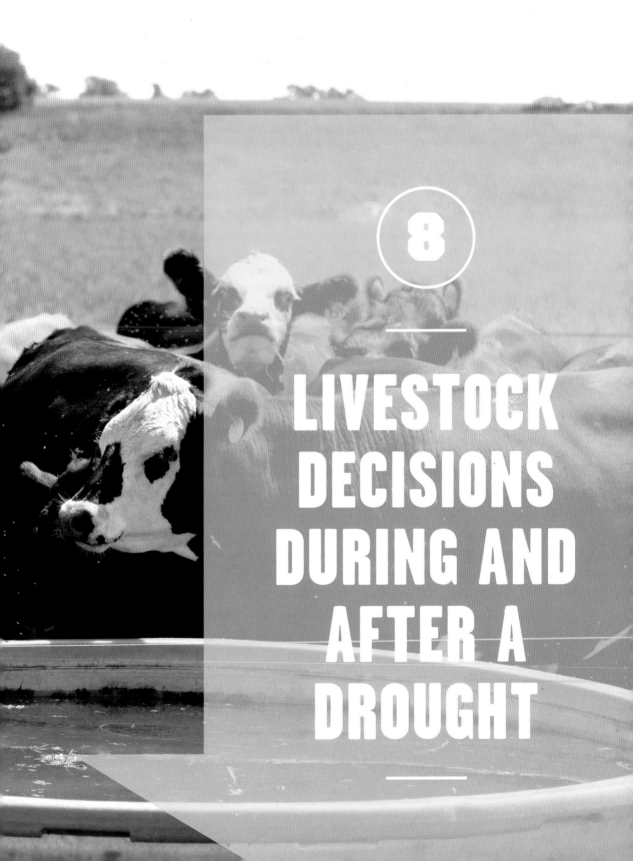

8

LIVESTOCK DECISIONS DURING AND AFTER A DROUGHT

The material we have covered so far involves preparatory steps you can take prior to a drought. This chapter deals with actions to be taken once the drought is in full force, to reduce its impact.

HAVE A PLAN

It is critical to have a written plan for dealing with drought. Drought is more than just a period of insufficient rain. It is a mentality. Far too often during a drought, out of panic or denial, we make poor decisions with long-term consequences. Having a plan on paper makes it more likely that you will stick to your plan when the event occurs, and it can eliminate the stress of trying to come up with ideas to get yourself out of the mess you are now in.

DECIDING WHETHER TO DESTOCK

Foremost among the decisions you need to make are whether it is necessary to reduce animal numbers. You must ask yourself several questions: Are my pastures below minimum heights? If they are, will there be any more pasture growth? Do I have enough stored feed on hand to feed livestock until pasture growth resumes after a rain? A concept that can aid the decision-making process is that of trigger dates (see box on facing page).

KEEP EASY-TO-DISPOSE-OF ANIMALS

One of the wisest ranchers I've ever met told me his primary drought strategy is to have half of each herd consist of **stocker animals** in addition to brood cows and calves. If it becomes obvious by his trigger date that a reduction in animal numbers is necessary, the first animals to go are the stockers.

Selling stockers is not a painful decision. They are meant to be sold. Going into the season knowing you will sooner or later sell these animals allows you the flexibility to sell sooner in the event of a drought. Selling cows in a drought, on the other hand, is a painful

Trigger Dates

Trigger dates are the dates on which decisions need to be made. For example, on warm-season native grass pastures in my area, good trigger dates are May 1, when you determine an initial stocking rate, and July 1, when half the growth of the grass will have occurred. On cool-season grasses, April 1 might be the first trigger date, and June 1 might be the second.

On the first trigger date, how you determine the initial stocking rate depends on your soil type. On loamy or clay soils you evaluate the amount of moisture received over the previous few months (left graph). With sandy soils you measure the depth of moisture (right graph).

The graphs below are from the USDA-NRCS in Kansas and are useful in making trigger date decisions.

Stocking Rate Reduction in Hardland (clay or loamy) Soils

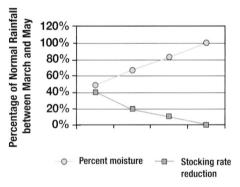

Table 8.1

Stocking Rate Reduction in Sandy Soils

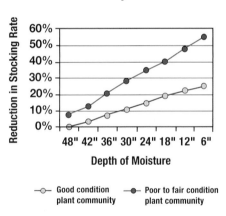

The second trigger date usually determines how long the grazing season will be. No matter how much it rains after July 1, the amount of grazing after this point will not exceed the amount that occurred previous to it. Nature has preprogrammed the grasses so that they produce most of their leaves by this time. If by this date your pastures are grazed below the recommended heights listed in chapter 6, Table 6.8, you must either find alternative feed or reduce animal numbers. Period. Whatever happens after this date will not change that fact, so be decisive and do it.

choice. Parting with cows is parting with future income. Besides, who is going to buy cows during a drought? No one else has the ability to feed them either. Cows sold during a desperation drought sale are most likely to be sold at rock-bottom prices.

This is actually a great time to expand a herd if you have the feed. Disposing of stockers in a drought and replacing them with otherwise good cows, which your out-of-feed neighbors are selling at rock-bottom prices because they can't afford to feed them, can be a way to rapidly build a cowherd without breaking the budget. Having a class of prede-termined animals that can be sold early in a drought is a fantastic tool.

KEEP GRAZING ANIMALS THAT DO NOT REQUIRE GRAZING

An alternative way to reduce grazing pressure during a drought without eliminating desirable animals is to have a class of animals that do not require pasture at all. Examples of this are pastured poultry or swine. If pasture growth is insufficient, you can confine the animals to buildings or dirt lots to reduce stress on pastures. Although pasture is quite beneficial to poultry and swine, it is not necessary. Of course, if you are marketing your animals as free-range or pasture-fed, complete removal from pasture is not possible, but you may be able to reduce the hours spent grazing per day to decrease pressure on the forage base.

MAKE A PRIORITY LIST FOR FIRST ANIMALS TO GO

Include in your drought plan which animals will be the first to go when a reduction is necessary. We have already pointed out the wisdom of having an easily disposed of class of livestock, but there are other animals that should be next on the list, or maybe even first.

The first glaring example might be cows that need culling. If a trigger date arrives and an animal reduction is necessary, one of your first actions should be a thorough animal assessment, including pregnancy testing. The criteria for removing an animal are often said to revolve around the "three Os": open, old, and ornery. Use the drought as an excuse to get rid of **open cows** (not pregnant), older animals, and those with behavior problems. Less obvious examples might be cows that do not fit your ideals for a cow, whether that is color or frame size.

Keep careful records of all sorts of information. For example, if you want to reduce fly issues on your cattle, having a record of cows that always have more fly problems makes it easy to dispose of those animals first if the need arises, and eliminates fly susceptibility genes from your herd.

HOW TO FEED YOUR WAY THROUGH A DROUGHT (AND HOW NOT TO)

If you have exhausted all the emergency grazing resources discussed in chapter 7 and you still need feed, and/or you are resist-ing destocking, it may be necessary to feed animals stored forage, such as hay, grain, or silage. There are correct ways and incorrect ways to do this.

First of all, don't wait until the pasture is overgrazed beyond repair. It is far better to feed a little hay while the pasture is still in good shape than to abuse the pasture and create a need for feeding hay even after rains return because the pasture is still not growing due to insufficient leaf area.

Second, don't leave the animals out on overgrazed pasture while you provide supplemental feeding. They will continue to graze every last leaf in preference to hay. It is better to have a sacrifice pen where you feed the animals hay while they are locked off pasture. This pen need not be a drylot; it can be a paddock growing a species that is tolerant of abusive grazing, such as bermudagrass, Kentucky bluegrass, or buffalograss. If you are in a rotational grazing system, you may be able to devise a method that allows the animals to graze for only a few hours a day, until target grazing heights are achieved, then remove them to a pen with hay.

Supplemental Feeding

In addition to hay, other feedstuffs can be useful as supplements. Grain can be a valuable help in stretching a pasture because it is more palatable than pasture; however, it is a very different feedstuff than pasture. Grain is primarily starch, while pasture is primarily cellulose. Starch ferments very rapidly in the rumen, while cellulose ferments very slowly. As discussed in chapter 7, too much starch at a time can cause acidosis, which can kill the microbes in the rumen and make animals completely unable to digest feed. It can even be fatal.

If you feed grain on pasture, start by offering it at a low level (no more than 0.5 percent of body weight daily) and slowly increase the amount. If you feed too much grain, the rumen microbial population shifts in favor of starch digesters and the cellulose digesters are killed off. Once this occurs, the affected animal will perform very poorly when placed back on pasture until the rumen readjusts, which may take as long as 6 weeks.

Grain is useful in stretching a pasture during the initial stages of a drought, but if it becomes apparent that the pasture supply is not sufficient, it's time to lock up the animals and feed hay or alternatives until the grass grows enough to support the next grazing episode. Acceptable alternatives to grain include distiller's grains, brewer's grains, cottonseed hulls, and soy hulls. These are feedstuffs with relatively low starch levels and thus can be fed at much higher amounts than grain without depressing the ability of rumen microbes to digest fiber.

Distiller's grains, for example, can comprise up to 25 percent of the dry matter intake without problems. In my area of the country, where there are ethanol plants everywhere, distiller's grains are cheap and abundant. Flaxseed, another option, has no starch, and its oil contains compounds that boost the immune system and increase the omega-3 fatty acid content of the meat.

SHIFT FROM GROWING-SEASON GRAZING TO DORMANT-SEASON GRAZING

If it becomes obvious that there are too many animals for the available forage, it can be beneficial to completely remove the animals from the pasture until dormancy has set in, after which the pasture can again be grazed completely without harm. In the meantime, animals can be fed hay, cover crops, failed grain crops, or any of the other emergency forages discussed previously.

What you want to avoid is the usual course of action, which is to graze the pasture until it is completely gone, then begin feeding hay prematurely out of necessity. This results in very little pasture regrowth, with extremely poor growth the following year, and the hay supply is usually used up before the winter is over. By contrast, complete pasture rest allows the grass to accumulate to the maximum possible with the given weather conditions, and this can be grazed during the winter, the normal hay feeding period. If you have 6 months of feed on hand, feeding it early (and sparing the pasture from abuse) can give you much better results than delaying feeding. This technique can keep pastures healthy during severe drought, maximizing the amount of feed available.

Bear in mind that forage left ungrazed during the growing season and grazed during dormancy does not have the quality of growing-season forage. Nor will hay support performance the way growing-season forage would, assuming an adequate quantity of both. Therefore, animal performance will be less optimal than with typical grazing but better than might occur with abusive grazing to bare dirt.

TIME-LIMIT GRAZING

Time-limit grazing represents a compromise between grazing continuously during a drought and switching to dormant-season grazing. The idea is to let animals graze only in the morning or only in the evening — reducing pressure on the pasture — and to feed hay in confinement when they are off the pasture. Unfortunately, this requires being present twice a day, once to open the gate to the pasture and once to chase the animals off the pasture, which can be difficult for someone with a full-time job. Time-limit grazing can be much easier with the use of a device called a **Batt-Latch**. This is a gate latch attached to an electrified slinky or bungee cord that acts as a gate. The latch releases at a preset time, and the gate springs open. This setup allows the rancher to be absent and still have the gate opened at a specific time.

Another method of time-limit grazing is to use a fish trap gate (see page 94). Animals feeding in the pasture will return to water via the fish trap gate, but once they do they cannot find their way back into the pasture. Hay is fed in the pen with the water source.

A Batt-Latch is a device that will open a gate automatically at a predetermined time, even if no one is present.

Time-limit grazing results in animal performance nearly equal to full grazing, since the high-quality pasture is blended with the lower-quality hay on a consistent daily basis.

FEED-STRETCHING ADDITIVES

Ionophores, like monensin (tradename Rumensin) and lasalocid (tradename Bovatec), alter the rumen fermentation so that feed is fermented more efficiently, and animals need about 10 percent less feed to perform the same. This is a well-documented phenomenon and can be a significant feed-stretcher. (See pages 119 and 125 for more information on Rumensin research.)

Feeding biochar can also have a beneficial effect (see pages 62–64). Reportedly, biochar adsorbs methane and keeps it from being belched out. The rumen microbes can use the methane as an energy source, converting it into microbial tissue that is digested by the animal.

How Biochar Affects Rumen Methane Emissions

	TOTAL METHANE EMITTED IN 24 HR (ML)	% DECREASE FROM CONTROL
Control	130	NA
1% biochar in diet	117	10%

Table 8.2

How Biochar Affects Weight and Feed Efficiency of Cattle

This research into liveweight gains was conducted in Laos at Souphanouvang University, and the study cattle were fed a basal diet of cassava chips.

	CONTROL	BIOCHAR AT 0.6% OF DIET	IMPROVEMENT
Average daily gain (lb)	0.23	0.28	22%
Feed efficiency (lb feed/ lb gain)	23.2	19.1	22%

Table 8.3

Alternate-Day Grazing

A twist on time-limit grazing is to graze and feed hay on alternating days. Animal performance is not as good as with time-limit grazing, due to inconsistent intake of quality forage, but this method can be easier to manage for the part-time rancher. Animals usually eat in the morning and again in late afternoon. Moving them on and off pastures at noon each day gives more consistent performance, since they will consume hay in one feeding per day and pasture in the other.

PERHAPS THE ABSOLUTE BEST THING YOU CAN DO IN A DROUGHT

Observe carefully and keep notes. Pay attention to what is happening and record it. Information allows you to plan more quickly the next time. With the right information, maybe you will be able predict when you will run out of pasture, so you can obtain hay before your neighbors begin to panic and the price of hay skyrockets. Maybe you will be able to market disposable livestock before the price goes down. Maybe you will remember to deepen that pond that went dry in that one pasture, or develop an alternate water source. Knowledge is power.

Here are some examples of what might be helpful to record during a drought:

- Note if the grass in a certain pasture died out earlier than others.

- Note when the pond dries up, or when a well gives out.

- Note if a weed problem moved in where it did not exist before.

- Keep track of how many head you had to reduce to per pasture.

- Keep track of how much supplemental feed you had to supply to each pasture.

- Create a list of "to do" items for the next drought. This may seem unnecessary during the drought when memories are vivid, but memories fade over time.

- Keep rainfall records and soil moisture observations.

- Record hay yields and grain yields.

AFTER THE DROUGHT

Here is a big question. Do you need to reduce stocking rates on pastures after a drought has passed?

The answer depends on how much damage occurred to the pasture species. If you put in place all the recommended procedures and precautions discussed in this book, your pastures just might come through with no damage and with good root reserves and adequate soil cover. If this is the case, no reduction in stocking rate will be necessary.

If, however, pastures were overgrazed, root reserves compromised, and the stand thinned, you may need to reduce stocking rates. It is often said that droughts have long tails: not because the drought itself lasts very long, but because the legacy of overgrazing can last for years. If overgrazing is prevented in a drought, life can return to normal the following year.

CHAPTER SUMMARY

- Have a written plan of actions to take during the drought.

- Identify trigger dates for your pasture types and your area. Typical trigger dates are the date of initial turnout to pasture, and the date by which 50 percent of the annual average pasture production occurs.

- Keep a portion of your herd as easily marketed animals, such as stocker steers, that can be the first animals to go when forage supply becomes low.

- Consider having animals that can maintain performance even if completely removed from pasture, such as swine and poultry.

- Prioritize animals in the order they should be marketed.

- Grain can be used to supplement pastures in the beginning stages of a drought to reduce grazing pressure, but too much grain (over 0.5 percent of body weight daily) can lower rumen pH and reduce forage availability.

- Hay is less palatable than pasture and seldom reduces grazing pressure. If hay is fed to stretch pasture supply during the growing season, it should be fed in a confined area off the pasture.

- If pastures begin to become grazed too low, it is better to remove animals completely and feed hay; the forage that is still in the pasture can be grazed after dormancy without harm.

- Time-limit grazing or alternate-day grazing can be used to maintain animal performance when most of the diet is composed of hay fed in confinement.

- Ionophores can increase feed efficiency and maintain animal performance on less pasture intake.

- Feeding biochar may also have merit to maintain animal performance on limited pasture, if legalities of feeding it are overcome by the next drought.

- Most importantly, keep a journal of your observations during drought. These may be invaluable later.

Any attempt to fight nature will eventually result in failure. The most successful agriculture systems will be those that mimic rather than attempt to conquer nature. One region where farming seems to experience a higher rate of crop failure than most is the Great Plains of the United States. For myriad reasons, this area has been largely converted to a system of farming wholly unsuited to the climate, and one that has degraded the ability of soil to produce in the future. Therefore, this region is one focus of particular attention in this book.

The other focus is your own farm or ranch. To help organize and guide your plan, I have included a checklist with which you can prepare your response to the next drought you yourself face.

PART III

LOOKING TO THE FUTURE

9

MOISTURE-EFFICIENT AGRICULTURE FOR SEMI-ARID REGIONS

The Dust Bowl of the 1930s has been called the greatest man-made environmental disaster in history. It occurred when a long-term drought exacerbated the problems related to the farming practices of the southern Great Plains region. Many of the same poor farming techniques were practiced over much of the world, but the climate of the Great Plains made the region particularly vulnerable to the results.

RECIPE FOR A "NATURAL" DISASTER

The first major mistakes that led to the Dust Bowl began almost 70 years earlier with the Homestead Act of 1862. Under the act, a family could receive 160 acres of land with the following stipulations: they had to construct a dwelling on the land of at least 14 by 12 feet in size, they had to "improve" the land by plowing it, and they had to live there for 5 years. While 160 acres of land seemed like a generous amount to someone sitting in a government office in the rain-fed eastern United States, it was wholly inadequate to make a living in the dry Great Plains, and about half the homesteaders abandoned or sold their claims within a generation.

The requirement of tillage resulted in the conversion of many marginal grounds unsuitable for cropping, lands that should have been left growing the native drought-resistant grasses. Even if the land had remained in native grass, 160 acres would have been enough to carry only about eight cows, given the standard "20 acres of grass per cow" rule of the shortgrass region — hardly a ranch of sufficient scale to support a family.

Since ranching on that small scale was not viable even then, plowing the sod for "better use" seemed at the time to be a wise choice. There was also in those days a prevailing notion (first proposed by climatologist Cyril Thomas and widely disseminated by author Charles Dana Wilber) that "rain follows the plow," based on the theory that tillage caused changes in the atmosphere that resulted in increased rainfall. Railroads and land speculators widely promoted this notion, telling prospective settlers that the parched Plains region could become as productive as the eastern United States, if only it were plowed.

It was soon discovered, however, that the highly variable and typically scant precipitation of the area, coupled with the region's hot, drying winds, resulted in frequent crop failures. Corn, for example, the crop of choice in the East, was more often than not a miserable failure in the West. Wheat, particularly the Turkey Red variety imported by settlers from Eastern Europe, proved to be more dependable than corn in the Great Plains. The peak water demand period (grain fill period) for corn occurs during July and August, typically the hottest two months of the year and generally the period with the least soil moisture of the year. The grain fill of wheat occurs in May and June, when rainfall is usually more abundant and heat is moderate. Yet even wheat failed in enough years that growing a crop every year was a risk.

DEAD-END SOLUTIONS

As settlers struggled to survive in the years following the Homestead Act, numerous methods were attempted to make farming viable in the area.

DUST MULCH

One of the more ridiculous (in hindsight) solutions was based on the so-called Dust Mulch theory, which proposed that the soil surface be finely tilled to break the capillary rise of water from the subsoil. This theory was first proposed in 1905 by a Dr. King, a University of Wisconsin agronomist, who observed that a column of soil in a glass cylinder would absorb water to a height of several feet when placed in a glass dish full of water. This led to the assumption that a similar phenomenon occurred in soil profiles, and that intense tillage would break the capillary contact between the subsoil and the surface and therefore conserve water.

In reality, intense tillage not only dried the soil but actually destroyed soil structure and decreased soil organic matter. This made the soil surface less conducive to water infiltration and more prone to runoff as well as erosion by wind and water. Although the Dust Mulch theory was rapidly discredited by sound science, it became widely adopted across the Plains and greatly contributed to the creation of the Dust Bowl.

DEEP TILLAGE

Another practice intended to improve soil moisture supply was deep tillage, or subsoiling. The idea was that if the soil were loosened, the open spaces in the soil would quickly fill with water during every rain. They would also store additional moisture and promote deeper rooting depth to extract the water. It was a compelling argument that seems to make sense, but scientific research showed this practice to be counterproductive. Numerous trials showed that deeply tilled soils actually promoted less infiltration, stored less water, and produced more shallow roots than soils with less tillage.

Despite the scientific findings, the practice continued. The destruction of soil aggregates and the reduction of organic matter from intense, deep tillage caused soil structure to break down. The loose structure initially created by deep tillage rapidly became severely compacted with each subsequent rainfall and with vehicle traffic.

As Kansas State Agriculture College extension bulletin 71 concluded in 1897, "Our experience with subsoiling, both for wheat and corn (Bulletin 64), indicates there is a positive loss, not only of labour but of yield. . . ." In the multi-year experiment reported in this bulletin, subsoiled plots yielded only 82 percent as much as the plots not subsoiled.

Another bulletin from Kansas State concluded in 1954 that in more than 60 years of research the authors had yet to find an advantage to subsoiling in the Plains and recommended the practice be discontinued. We have known for more than a century that subsoiling is a dead end, and yet it persists as a popular practice today. (For more on subsoiling, see chapter 4.)

FALLOW

Another practice that came into use about the same time was that of **fallow**. The idea was to grow a crop only every other year, and to use the off-year to store soil moisture. Unlike the Dust Mulch and Deep Tillage theories, this one did reduce the rate of crop failure. It was horribly inefficient, however, in that only about 10 percent of the moisture received during the fallow year was still present at the planting of the next crop. Even so, in many years this extra 2 inches of moisture at planting time made the difference between a crop failure and a success.

Fallow as a practice became more successful with the development of the wide-bladed sweep plow, or "V-blade," by Charles Noble of Alberta. Often called the Noble blade, it was patented in 1937, but its use became common in the 1950s and continued as a common practice through the 1990s.

This tillage tool killed weeds by cutting the roots without burying the crop residue. It left the residue on the surface as a mulch, which led to the practice of "stubble mulching." This made moisture storage in fallow much more efficient than clean, tilled "black fallow" and increased fallow efficiency to 15 to 20 percent. Leaving stubble on the surface also reduced soil erosion. There was

A Noble blade in use. Note how the crop residue remains on the soil surface to retain its value as mulch.

a drawback, however, in that the use of the Noble blade created a dense tillage pan that restricted root growth of subsequent crops.

In the 1960s and '70s the development of herbicides, particularly atrazine, paraquat, and glyphosate, made possible a system called **chem-fallow** (or eco-fallow), which involves fallowing wheat stubble with no-till and herbicides. This led to a further improvement in fallow efficiency, raising it to 25 percent. Chem-fallow not only improved moisture efficiency, but it also greatly reduced the amount of fuel usage, labor requirements, and soil erosion. It became the dominant system when glyphosate went off-patent (and therefore became cheap) in the late 1990s.

Questions and Concerns

While fallow led to a reduction in crop failures, it had some serious drawbacks. It is highly inefficient even in its most efficient form, which is no-till chem-fallow. Fallow "works" by storing 25 percent of the moisture from one year to add to the next year's in-crop precipitation, resulting in a yield similar to what you might have in a year with very favorable precipitation. But does this really increase the efficiency of moisture use?

To answer that question, let's assume we are in a region with 20 inches of annual precipitation. As mentioned, a fallow system will store, on average, 25 percent of that 20 inches, or 5 inches, and add it on to the next year's 20 inches. In other words, in year 1 you get 20 inches of rain and no crop. In year 2 you get the equivalent of 25 inches of rain and a single year's crop yield. So you've effectively converted 40 inches of rain into 25 inches of rain used by one crop. This is wise?

Another significant drawback of fallow is that it deprives soil microorganisms, including mycorrhizal fungi, of vital root exudates for nearly a full year. This leads to local extirpation of mycorrhizal fungi and other rhizosphere organisms — the very same organisms that would otherwise greatly aid the water efficiency and drought tolerance of the crop.

This raises the question: Is fallow worth it? We went down this misguided route because we thought the holy grail of farming was grain, and fallow made grain production possible in a place where it was otherwise unsuccessful. Like Captain Ahab's reckless pursuit of Moby Dick, we destroyed entire ecosystems in our quest to produce grain in an environment that does not naturally produce large seed yields. The native grasses of the Great Plains have miserably poor seed yields, much less than the natural seed yields of the oak forests of the eastern United States, or even the tall grasses of the prairie region. Instead of seed, reproduction is primarily vegetative through stolons and rhizomes, a process that takes less moisture and relies less on timely rainfall during seed germination.

The production of grain is very water-intensive compared to the production of vegetative material, as evidenced by the graph on the next page showing the water use of wheat over its life cycle. Note the sharp increase in water demand after the jointing stage, which is when the reproductive stage begins.

Water-Use Curve of Wheat

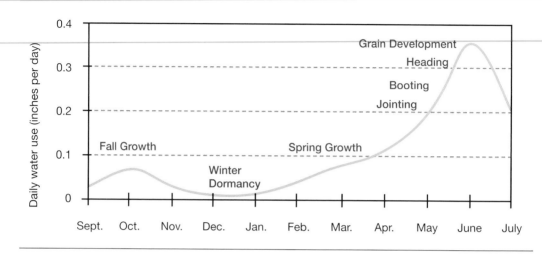

Table 9.1

The Drought-Resilient Farm

A BETTER DIRECTION

Perhaps we would better serve ourselves by abandoning our single-minded pursuit of grain in semi-arid regions, and instead transition to a form of agriculture better suited to the climate: that is, growing vegetative material that is harvested by ruminant animals, just as buffalograss and blue grama produced leaves harvested by bison for centuries. This is not to propose the creation of a "Buffalo Commons" as Rutgers professors Frank and Jane Popper famously proposed in 1987 (see box on page 165); rather, it offers a much more profitable use of the limited and unpredictable moisture of the region. This method of farming would have six key components.

1. REPLACE FALLOW WITH GRAZEABLE COVER CROPS.

Many "experts" claim that cover crops do not belong in arid regions, that the moisture they use is not reliably replaced by subsequent rainfall, and that they often harm the yield of the following crop. This is indeed what short-term research would indicate. Grain crops following cover crops usually show a small yield disadvantage compared to the same crops following fallow.

There is another side, however, to this research. Cover crops in these trials often produced as much as 3 tons an acre of biomass, which in the real world could have been harvested by ruminant animals. In the drought year of 2012, how valuable would 3 tons of forage have been? Far more valuable than the four-to-seven-bushel decrease in the subsequent grain crop, during the driest year in history. Hay in that year was bringing over $200 a ton. While pasture is not priced as high as hay because it is not transportable like hay, a ton of pasture can eliminate the need for a ton of hay.

It is during drought years that a cover crop grown on moisture received prior to the drought can be most valuable. For example, the most productive cover crop I ever had was the one grown over the warm, open winter of 2011 to 2012. It was also my most profitable because the pasture it provided was invaluable when the rain shut off that summer.

It was in 2012 that I became convinced that cover crops were a good idea in semi-arid regions. I observed that (in fields often adjacent to each other):

- corn in tilled soil completely failed,

- corn no-tilled into soybean stubble produced a small yield,

- corn no-tilled into wheat stubble produced a moderate yield,

- and corn no-tilled in land that had been cover-cropped for many years produced a normal yield.

Here is the logic behind using grazeable cover crops in the Great Plains: Since moisture storage is only 25 percent efficient, don't try to store moisture, but rather use it as it falls. (Could you imagine going in to get your paycheck and being asked, "Do you want 100 percent of that now, or would you rather have 25 percent of that a year from now?") Convert rainfall into biomass as it falls, and convert part of this biomass into meat or milk, and leave the other part of the biomass as mulch.

A growing body of evidence also indicates that grazed cover crops have a much more beneficial effect on soil characteristics than ungrazed cover crops do. Obviously, cover crops, grazed or ungrazed, also have the strong residual effects of promoting infiltration (chapter 2), increasing organic matter and reducing evaporation (chapter 3), and improving rooting depth (chapter 4), in addition to providing forage at critical times (chapter 8). Cover crops used for grazing can serve as fallow replacements in an existing grain crop rotation, or the cropland can be dedicated to grazing crops only.

Alternatively, sorghums that do not produce grain (male sterile or

Confused? Picture It in Dollars

One way to think about this is to compare the conversion of moisture into dollars.

- An inch of moisture will raise about 11 bushels of corn grain.

- It takes 4 inches of fallow rainfall to create 1 inch of stored soil moisture for next year's corn.

- That 4 inches of rain can grow up to 800 pounds of biomass per inch with a cover crop of sorghum-sudan, for a total of 3,200 pounds.

- Graze 40 percent of this, and with a 10:1 conversion ratio of green pasture to beef gain, that cover crop can produce 128 pounds of beef.

- Compare the value of 11 bushels of corn to 128 pounds of beef, and you tell me which is worth more.

photoperiod-sensitive) can be planted in wheat stubble and stockpiled for use as a very high-yield winter pasture. A successful pair of ranchers I know near the western Nebraska town of Culbertson, father and son duo John and Jacob Miller, use stockpiled cover-crop blends based on sorghum as the entire wintering program for their cowherd. They feed hay only during the occasional blizzard, more out of sympathy than necessity, they say, as the tall sorghum is easily accessible even in deep snow. During the winter of 2015 to 2016, they wintered 750 animals with only 6 bales of hay.

If you take a calendar and plot out the months of grazing that cover crops could potentially offer in this very common crop rotation, you will find that grazing is provided in essentially every month. During any time of shortage, you can simply move the livestock back onto pasture.

2. MAINTAIN FLEXIBILITY.

One of the more innovative ideas for semi-arid cropping systems is the concept of flexible cropping rotations, developed by the USDA Agriculture Research Service (ARS) at Akron, Colorado. It is based on the idea that the amount of stored soil moisture in the ground at planting time plays a large role in the success of the crop. The procedure is to measure the soil moisture at each planting opportunity, then add an assumed 75 percent of average rainfall for the growing period of the crop. You then compare this amount to a chart that shows how much moisture is needed to profitably grow each of a variety of crop choices.

For example, assume you have land not currently planted, and it is April. If there is a full soil profile, the most profitable choice is corn. With a little less moisture, grain sorghum

Splicing Grazed Cover Crops into Your Rotation

Here is an example of how to add grazed cover crops into a Great Plains crop rotation.

A typical dryland crop rotation would be corn, followed by a year of fallow after harvest, then a crop of winter wheat, followed by almost another year of fallow before corn is planted the following spring. This rotation provides two opportunities for cover crops during each fallow period.

Soon after corn harvest in fall, plant a cover-crop blend based on a winter cereal or winter legume; or, in early spring, plant a cover-crop blend based on oats and peas. The winter cover crops could be grazed in early spring, the spring-planted cover crops in late spring or early summer.

After wheat harvest, immediately plant a blend of summer annuals, such as sudangrass, millet, and cowpeas, and use it for late-summer grazing; or, in late summer, plant a blend of both summer annuals and cool-season crops, such as turnips, radishes, peas, and oats, and use that for fall grazing.

is most profitable. With less moisture yet, millet grain or sorghum-sudangrass pasture is most profitable. If it is too dry to get a crop to germinate, simply wait until the next rain. This approach has been more profitable than any rigid crop rotation in the USDA-ARS projections during this long-term study. See References for more information.

3. USE NATIVE SHORTGRASS AS FLEX PASTURE.

I listened to a rancher from eastern Colorado describe his pasture management system at a conference many years ago. One of the audience members, a rancher from the Corn Belt, stood up and said, "You have it so easy in your area." This took the speaker by surprise, since the general perception is that ranching is much tougher in a 12-inch rainfall area (eastern Colorado) than in a 36-inch rainfall area (Corn Belt). He replied to the audience member, "I am confused. What do you mean by that?"

The Corn Belt rancher then explained himself. "You have a grass [buffalograss] that keeps its quality 12 months a year. You can even grow it one year and graze it the next and it will still be good. Our pasture needs to be grazed as it grows — if it gets ahead of the livestock, it turns woody and the animals won't eat it. We must have constant moisture to grow our pastures. Yours can survive on intermittent moisture and thrive."

The audience sat there stunned. The Corn Belt rancher had pointed out a big advantage to the grasses of the shortgrass plains that is seldom capitalized on: they maintain quality at all growth stages, and that quality persists for a long time after they have grown.

Allan Nation, the late editor of *The Stockman Grass Farmer* magazine, said that the key to a successful agriculture enterprise is to find the advantage your area has over other regions of the country and capitalize on it. The "never gets woody" nature of short-grasses like buffalograss and blue grama is the true advantage of the Great Plains.

Capitalize on this feature by using short-grass as "flex pasture." During much of the year, it will be possible and desirable to graze cover crops, as explained above. During times when cover crop grazing is not available, simply move over to the buffalograss and resume grazing. It will still be good no matter when you get around to it, and the respite it has received in the meantime will allow it to be nice and healthy, with a full root system.

4. TURN DROUGHT-FAILED CROPS INTO A PASTURE RESOURCE.

Drought can ruin the best-laid pasture plans. Typically, when drought occurs, the first stage of pasture management is denial: "It will rain soon." Therefore, action is delayed, and the pastures get grazed too short. This causes a cascade of negative effects. Litter cover is reduced, so evaporation increases and rainfall infiltration decreases. Carbohydrate storage in the roots is reduced, so regrowth slows, and if this persists the long-term negative effects can last for months.

Once the pasture is grazed to the recommended minimum stubble height, it is imperative to remove the animals to an alternative feed source. The problem during a drought is that we seldom have alternative feed sources. Hay obtained at very high cost is the usual route, and feeding your way through a drought with hay can be an extremely expensive process.

One emergency feedstuff few people think of is the grain crop adjacent to the pasture, which may or may not make a grain crop in a drought year. (This option is described in detail in the first section of chapter 7.) In the Great Plains, this might be not just an emergency plan but also an option that can be utilized anytime pasture gets short.

It is not necessary to graze the entire corn or wheat crop. With effective temporary fencing, you can begin strip grazing at one end, grazing only the amount of area necessary for that day, one day at a time. Then you can plan on harvesting the rest for grain if a crop-saving rain occurs at some point or if there is crop left over at the end of the grazing period.

5. USE SUCCESSFUL GRAIN CROPS AS FATTENING PASTURE.

If there is enough rainfall to make a good grain crop, why not harvest it by strip grazing with feeder calves rather than with a combine? I fail to understand why this is perceived as such a radical idea. Isn't most of the corn and sorghum produced in the Great Plains used to feed cattle anyway?

Think of it this way. The "normal" way to farm is to have calves in a pasture and corn in a field adjacent to that pasture. In the fall we round up the calves and put them on a $100,000 truck that runs on $2.50-per-gallon diesel fuel, and we ship those calves to a

feedlot. Then, we fire up a $350,000 combine (using more diesel fuel) to get the corn out of the field, put it on a $100,000 truck (more fuel), and ship it to a multi-million-dollar grain elevator, where we get docked for excess moisture. The elevator burns propane to dry the grain for safe storage, then loads it on a multi-million-dollar train (which also runs on diesel) and ships it to the same feedlot where the calves were previously sent. Those calves turn the grain into money and manure. Because manure is bulky and expensive to transport, the manure is spread on adjacent fields at levels high enough to cause actual yield reductions on crops due to excess salt content. Meanwhile, back at the farm, the farmer uses much of his income from the corn crop to buy fertilizer to replace the fertility that was contained in the corn he trucked off.

The "weird" way to farm is to replace that process by opening the gate between the pasture and the corn field and spending 10 minutes a day of pleasant exercise using polywire to strip-graze the corn after it is mature. The corn gets changed into higher-value beef, and the manure enriches the soil and reduces next year's fertilizer bill. It takes approximately 55 bushels of corn to change a 750-pound feeder calf into a 1,350-pound fat calf.

If you want to pencil out the economics of this, take the value of a fat calf, subtract the value of a feeder, and divide that difference by 55 to get what you are selling your corn for on the hoof. I guarantee you will be pleasantly surprised. You can also deduct the harvest cost you saved and add back the value of the fertilizer you retained. "Weird" farming looks pretty good compared to "normal" farming.

6. PUT THE LAND BACK TO PERENNIAL GRASSES.

This is not advocating a "Buffalo Commons" (see below). The idea here is not to replant just buffalograss and blue grama, the current native vegetation. Instead, plant grasses like eastern gamagrass or big bluestem that are more productive on deeper soils or where irrigation is limited.

Buffalo Commons

In 1987, Rutgers University geographers Frank and Deborah Popper published an essay in *Planning* magazine titled "The Great Plains: From Dust to Dust" (see References). They pointed out that since 1920 the Great Plains had experienced a population loss of roughly a third (a trend that continues to this day), and that across this region there are more than 6,000 ghost towns. They proposed the creation of a huge national park encompassing much of ten states stretching from eastern Montana down to west Texas, in which the natural Plains vegetation would be recreated and the original wildlife, primarily bison, restored.

This concept became known as "The Buffalo Commons." It was widely ridiculed and derided when it was first publicized, but the Poppers' essay did point out what too many people have denied for too long: the current model of agriculture in the Great Plains is not working for the ecosystem, the economy, or the people of the region.

Prior to the Dust Bowl these grasses were naturally abundant, but they could not handle the intensive grazing that took place during the Dust Bowl and the later drought in the 1950s. They were replaced with the less productive but more grazing-tolerant buffalograss and blue grama. Since they are seldom found in pastures in the area now, it is often mistakenly assumed they do not thrive in the area.

In addition to the grasses, include legumes (such as alfalfa) to provide nitrogen to the mixture as well as additional protein, and plant deep-rooted forbs, like Maximilian sunflower and bush morning glory, which can access moisture from depths that most grass roots cannot.

The key to success with these plantings is diversity for added resilience. For example, although the bulk of the mixture should focus on more productive (and water-demanding) species, it should also include a sprinkling of buffalograss and western wheatgrass. These species will survive if the worst-case scenario (severe drought coupled with severe overgrazing) ever occurs again.

IMPROVING IRRIGATION

Shortly after World War II it seemed that the problem of drought was solved on much of the Great Plains when the discovery of the underground water supply in the deep Ogallala and High Plains Aquifers coincided with the development of the center-pivot irrigation system. Freed from the limitations of natural rainfall, farmers soon found that their land could surpass the productivity of the Corn Belt in corn yield per acre. A look through the window of an airliner flying over the region reveals acre after acre decorated with the green circles on a brown background created by center pivots in an otherwise arid landscape.

By the 1960s, the availability of huge quantities of corn, the distance from terminal markets that depressed grain prices, and the arid climate that made mud a rarity all contributed to making this region a mecca for fattening cattle on grain. This led to the development of the feedlot industry, followed soon thereafter by the packing plant industry. Refrigerated trucks and rail cars made the shipment of refrigerated meat in boxes cheaper than the transport of live cattle to slaughter in distant population centers, so the meat-packing industry moved closer to the feedlots. All this combined to make many of the farmers who survived the Dust Bowl quite wealthy and the entire area prosperous.

Today, however, this prosperity is gravely threatened. In my native Kansas, the Ogallala Aquifer, which underlies the western third of the state, is being rapidly depleted by irrigation at levels far exceeding a sustainable rate. Already, many irrigation wells have been turned off because it costs more to pump a cubic foot of water than the value of grain created by a cubic foot of water. Most other wells have experienced a serious decline in productivity, and the future of irrigated agriculture in the region is in jeopardy, along with the economy that depends on it. The status of the aquifer is shown in the illustration on page 168.

To make irrigation sustainable we must reduce the amount of water we apply annually to a level comparable to the natural replenishment rate. In many areas, this amount is considered to be 4 acre-inches. Given that

An eagle's view of center-pivot irrigation

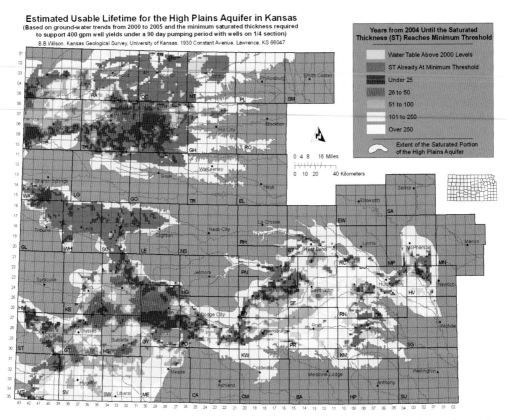

Estimated Usable Lifetime for the High Plains Aquifer in Kansas
(Based on ground-water trends from 2000 to 2005 and the minimum saturated thickness required
to support 400 gpm well yields under a 90 day pumping period with wells on 1/4 section)
B.B Wilson. Kansas Geological Survey, University of Kansas. 1930 Constant Avenue. Lawrence. KS 66047

**Years from 2004 Until the Saturated
Thickness (ST) Reaches Minimum Threshold**

- Water Table Above 2000 Levels
- ST Already At Minimum Threshold
- Under 25
- 26 to 50
- 51 to 100
- 101 to 250
- Over 250
- Extent of the Saturated Portion of the High Plains Aquifer

0 4 8   16 Miles

0 10 20   40 Kilometers

In many areas the Ogallala Aquifer is being rapidly depleted. This map shows its estimated usable lifetime at current use rate in Kansas.

an irrigated corn crop requires 26 acre-inches of growing-season moisture from any combination of rainfall or sprinkler irrigation to maximize yield (over most of the Great Plains), this reduction is likely to occur only where the total growing season moisture exceeds 22 inches. This assumes, however, that we are following current irrigation efficiency standards. It is possible to dramatically reduce the amount of water needed by improving irrigation efficiency.

There are several ways to improve irrigation efficiency over the standard center-pivot systems currently in use. With standard systems, too much water evaporates before it hits the soil surface, too much runs off the surface before it infiltrates, and too much

evaporates from the soil surface after it does infiltrate. The following methods can be used to reduce those losses.

Use drop nozzles rather than overhead nozzles. The closer the water is released to the ground, the lower the evaporation. Ideally, water should be released right above the crop canopy. If released below it, the spread pattern is disrupted.

Utilize drag-behind drip lines rather than nozzles to release water right on the soil surface.

Plant a ring of trees as a windbreak at the outside of the pivot pattern, and use the pivot end gun to irrigate the trees during establishment. Trees can be selected to produce a useful product, such as fruit, nuts, or lumber, in addition to slowing the wind and reducing evaporation.

Apply all the techniques described in chapter 2 to promote rapid water infiltration. This would allow you to slow down the pivot so that it applies more water at a time, without runoff. I will never understand why anyone chooses to put a tillage implement in the ground in the semi-arid High Plains. I hear there is a problem with too much residue to plant through on no-tilled fields, especially continuous corn. To me, that is a sign of a grazing opportunity. "Too much residue" is an asset, not a problem, particularly in drier areas.

ALTERNATIVES TO CENTER-PIVOT IRRIGATION

A far more efficient form of irrigation than center-pivot is subsurface drip tape. Drip tape offers many advantages over pivot irrigation:

- No losses due to evaporation or runoff because the water is released below the soil surface

- Rainfall is more likely to be stored, even on a freshly irrigated field, because the soil surface remains dry during irrigation

- Soil is often cracked during dry weather, even while it is being irrigated, providing three benefits: rapid infiltration of rainfall, natural alleviation of compaction, and oxygen penetration into the deeper soil regions; drip tape allows soil cracking without crop stress

- Requires far less pumping power than center pivots

- Much less likely to be destroyed during tornados than pivots

Drip tape is economical and efficient for watering long rows. Inside the tape are spaced emitters that release a slow and consistent flow of water.

- Capable of delivering nitrogen fertilizer right to the root zone, reducing the risk of runoff

- Adaptable to many field shapes unsuited to pivots

- Does not leave pivot tracks where gullies form

- Does not require specialized fence crossings if subdivided for use as pasture

- Uses only 50 to 75 percent of the water that a center pivot does, typically

An even more efficient method than drip tape is an ancient form of irrigation called **ollas**. Ollas are simple unglazed clay pots that are buried in the ground and filled with water, and seeds are planted around them. The simple reason for their efficiency is that water is released from the pots only when the soil around them becomes dry, and not before. Ollas work on the osmotic pull that dry soil exerts on the water in the pot, so water moves toward growing roots as they dry the soil. The disadvantage of ollas is that they are difficult to utilize on a large scale, as each one is typically installed and filled by hand.

One potential adaptation of the principle behind ollas is the use of a drip tape composed of a semipermeable membrane. This could combine the water use efficiency of ollas with the quick installation and lower expense of drip tape for large-scale applications. One such technology was briefly on the market, under the trade name Root Demand Irrigation, but it was scuttled due to legal issues. This tape showed tremendous promise, releasing water only when and where it was needed. According to the limited information available, it reduced irrigation water applied by as much as 40 percent compared to traditional drip tape, without any reduction in crop yields. It could also operate under much lower pressures than other irrigation systems (4 psi compared to the 15 psi needed for drip tape or the 40 psi for a pivot), significantly reducing pumping requirements and energy consumption. If and when the legal issues of Root Demand Irrigation are resolved, this technology could dramatically prolong the life of irrigation in the High Plains.

Ollas are among the most efficient small-scale irrigation systems known.

CHAPTER SUMMARY

Much of the semi-arid areas receiving between 10 and 20 inches of moisture a year are operated under horribly moisture-inefficient agriculture systems. You can improve the efficiency of these systems in many ways:

- Replace fallow with grazed cover crops.

- Replace fixed crop rotations with flexible rotations based on available soil moisture.

- Utilize native buffalograss and blue grama pastures as "flex pastures" to be grazed when cover crop grazing is unavailable, and rested while cover crops are being grazed. This rested pasture can be accumulated for later use as needed.

- Turn drought-stricken crops into pasture.

- Utilize successful grain crops as a fattening pasture through strip grazing after grain maturity.

- Restore land to improved blends of grasses, legumes, and forbs to create a highly drought-tolerant system and add valuable organic matter to soil. Perennial plantings can be used either as permanent plantings or in rotation with grain crops.

You can improve center-pivot irrigation efficiency by using:

- Drop nozzles

- Drag-behind drip lines

- Circular windbreaks at the edge of the end-gun pattern

- Soil management to increase infiltration and reduce runoff (methods described in chapter 2)

Subsurface drip tape offers many advantages over pivot irrigation, such as:

- No evaporation or runoff of applied water

- Reduced pumping costs

- No pivot tracks

- Natural compaction relief due to cracking of soil surface

- Adaptability to odd-shaped fields

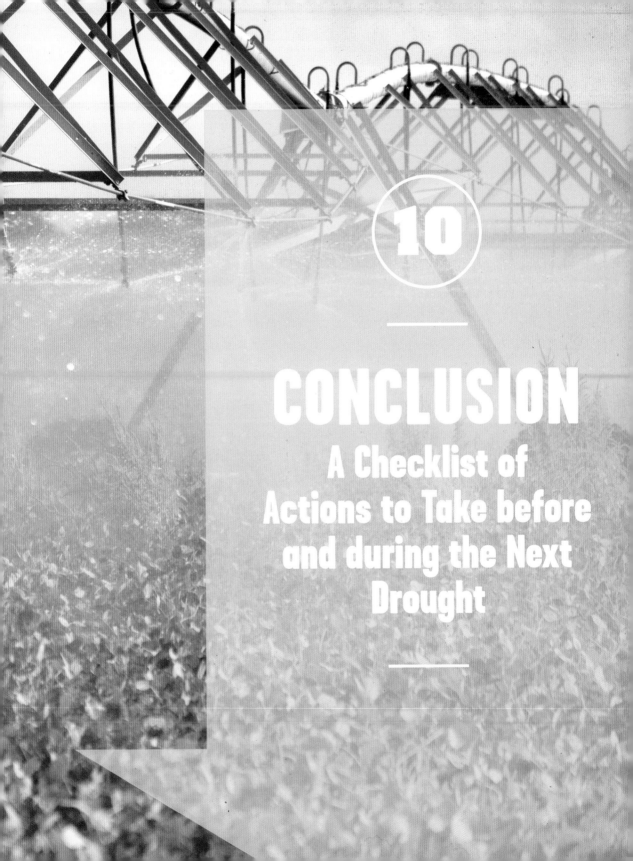

10

CONCLUSION

A Checklist of Actions to Take before and during the Next Drought

Drought has always been considered a phenomenon induced by nature. We have felt for centuries that our only possible course of action is to pray. Hopefully, after reading this book, it is apparent that there are hundreds of actions that can be taken prior to or during a drought to cushion its impact.

I thought it fitting to end this book with a checklist. Go down this checklist and check off the actions you have taken to prepare for drought, and cross out those that do not apply to you. Don't just sit on the rooftop and pray. Pray, and then act. Exhaust every action you can to prepare for the next drought, and exhaust every possible action during the drought.

IMPROVING THE SOIL

- ☐ Eliminate all tillage.
- ☐ Retain all crop residue.
- ☐ Create additional mulch with cover crops.
- ☐ Import additional organic materials as a mulch.
- ☐ Add manure.
- ☐ Inoculate with mycorrhizal fungi.
- ☐ Eliminate weeds.
- ☐ Add biochar to soil.
- ☐ Maintain optimal levels of soil fertility and pH.

FEEDING PRACTICES

- ☐ Utilize weeds as livestock feed.
- ☐ Feed hay on pastures and fields rather than in lots to disperse manure and refused hay.
- ☐ Feed biochar to livestock to supply biochar nutrients to pastures via manure.

MANAGING THE LANDSCAPE

- ☐ Form swales.
- ☐ Create vertical mulches.
- ☐ Create mini-dams along seasonal water courses in pastures.
- ☐ Retain, maintain, or establish windbreaks.
- ☐ Leave tall stubbles on both pastures and cropland.
- ☐ Create mini-ponds at points as high on the landscape as possible.
- ☐ Exclude livestock from water sources like ponds and streams.
- ☐ If you farm in a very drought-prone area, realize that forage production is more moisture-efficient than grain production.
- ☐ Convert inefficient irrigation systems, such as flood systems, to subsurface drip.
- ☐ On smaller acreages where hand labor is justified, install ollas.

MANAGING PASTURE

- [ ] Take steps to ensure that pastures have proper pH and fertility.

- [ ] Prioritize deep-rooted pasture species.

- [ ] Warm-season grass pastures will be less susceptible to summer drought than cool-season grass pastures. If the pasture base is cool-season species, consider interseeding warm-season annual grasses like corn, sorghum, millets, crabgrass, or teff.

- [ ] Never overgaze your animals, the single best measure to improve pasture drought tolerance.

- [ ] Avoid severe grazing when grasses are in the reproductive stage.

- [ ] If available at low cost, use wood chip mulch on pastures, particularly in high traffic areas prone to compaction and water runoff.

COPING DURING A DROUGHT

- [ ] Use failed crops for emergency pasture, hay, or silage.

- [ ] Utilize crop residue for pasture or hay.

- [ ] Ammoniate low-quality crop residue to increase its value.

- [ ] Utilize cover crops for grazing, especially during periods when perennial pastures are vulnerable to overgrazing.

- [ ] Cut down trees and use the leaves for forage.

- [ ] Graze dormant grass with protein supplement.

- [ ] Use feed-stretching supplements such as *Aspergillus oryzae* extract and ionophores like lasolocid and monensin.

- [ ] Consider using lawn clippings, if available, to stretch pastures. Be sure to utilize within 24 hours of cutting.

- [ ] Control grasshoppers to minimize competition with livestock for forage.

- [ ] Graze wetlands that should otherwise be kept free of livestock.

- [ ] Keep a percentage of your herd as easy-to-market animals such as steers.

- [ ] Have a written plan for trigger dates, upon which pastures are evaluated and destocking decisions made.

- [ ] Consider having livestock that can be pulled off pastures without sacrificing performance, such as poultry or swine.

- [ ] Keep a priority list of animals that will be first to go when destocking becomes necessary.

- [ ] Consider feeding grain byproduct feeds at low levels at the beginning of a drought to keep from overgrazing.

- [ ] Never, ever use supplemental feeding to keep animals on pasture after it is grazed too low.

- [ ] Have a pen or sacrifice area to feed emergency hay; never feed hay on pasture after minimum grazing heights are reached.

- [ ] Consider time-limited grazing schemes such as alternating morning grazing and evening hay, or grazing one day and hay feeding the next.

- [ ] If stored feed (such as hay) is limited, use it during grazing season to prevent overgrazing, and graze off stockpiled pasture during the dormant season to cause less damage.

- [ ] Feed biochar for improved feed efficiency and better nitrogen cycling.

- [ ] Keep careful written and photographic notes of observations before, during, and after a drought. These notes can be valuable later.

ACKNOWLEDGMENTS

We all stand on the shoulders of giants. No one pops out of the womb and writes a book without being helped, shaped, informed, or inspired by others. I almost hesitate to write this list, as I know I will omit so many who helped me on this journey. Some people to whom I am grateful for their help, directly or indirectly:

My brother Todd, who read over my manuscript and offered advice

Chad Buckley, the first person who ever told me, "You should write a book"

Stanton Gartin, who believed in me enough to offer me my first full-time job teaching agronomy at Cloud County Community College

All my professors in the Department of Agronomy at Kansas State University, who taught and inspired me, notably Stan Ehler, Steve Thien, Gerry Posler, and Charles Rice, among many others (if you remember my name from that far back, you probably belong on this list as well)

Dr. Clenton Owensby, who saw enough promise in me to offer me a graduate position

Dr. Mike Amaranthus and Larry Simpson, for photos and education about mycorrhizal fungi

Will Boyer and Herschel George, who provided photos of watering systems

Keith and Brian Berns of Green Cover Seed, who introduced me to so many inspirational and knowledgeable people, including the biggest names in the soil health movement, and allowed me to further my education on the job

Dan and Mary Howell, Ryan Carlgren, Shane New, and Jay Ronnebaum, who put up with all my phone calls and patiently listened to all my left-field ideas without audibly laughing at me

All my former students, customers, and co-workers who helped shape my ideas throughout the years

And all the others who encouraged me to put my ideas into print.

APPENDIX

REFERENCES AND FURTHER READING

CITATIONS

Tables

Table 2.1. Boyle et al. (1989). *Journal of Production Agriculture*, 2:4, pages 290–299.

Table 2.2. Mannering & Meyer (1963). The effects of various rates of surface mulch on infiltration and erosion. *Soil Science Society of America Proceedings* 27:84–86.

Table 2.3. Data are an average of multiple sources compiled by the author.

Table 2.4. Donahue, Follett, Tulloch (1976). *Our Soils and Their Management, Fourth Edition*. Interstate Publishers.

Table 2.5. [GRAPH] Blanco, Mikha, Presley, Claassen (2011). Addition of cover crops enhances no-till potential for improving soil physical properties. *Soil Science Society of America Journal* 75(4): 1471–1482.

Table 2.6. Donahue, Follett, Tulloch (1976). *Our Soils and Their Management, Fourth Edition*. Interstate Publishers.

Table 2.7. Lee (1985). *Earthworms: Their Ecology and Relationships with Soils and Land Use*. Academic Press (Elsevier, Inc.).

Table 2.8. Kladivko (1993). *Earthworms and Crop Management*. Purdue University Cooperative Extension Bulletin AY-279.

Table 2.9. Shipitalo, Protz, Tomlin (1988). Effect of diet on the feeding and casting activity of *Lumbricus terrestris* and *L. rubellus* in laboratory culture. *Soil Biology and Biochemistry*, 20:2.

Table 2.10. Blanco, Mikha, Presley, Claassen (2011). Addition of cover crops enhances no-till potential for improving soil physical properties. *Soil Science Society of America Journal* 75(4): 1471–1482.

Table 2.11. Kladivko (1993). *Earthworms and Crop Management*. Purdue University Cooperative Extension Bulletin AY-279.

Table 3.1. Shantz & Piemeisel (1927). The water requirements of plants at Akron, Colorado. *Journal of Agriculture Research* 34:1093–1190.

Table 3.2. Gifford, Bremner, Jones (1973). Assessing photosynthetic limitation to grain yield in a field crop. *Australian Journal of Agricultural Research* 24, 297–307.

Table 3.3. Radke & Burrows (1970). Soybean plant response to temporary field windbreaks. *Agronomy Journal* 62: 424–429.

Table 3.4. Aase & Siddoway (1976). Influence of tall wheatgrass barriers on soil drying. *Agronomy Journal* 68(4): 627–631.

Table 3.5. Top graph: Blanco-Canqui, et al. (2006). Corn stover impacts on near-surface soil properties of no-till corn in Ohio. *Soil Science Society of America Journal* 70: 266–278.

Bottom graph: Nielsen, et al. (2005). Efficient water use in dryland cropping systems in the Great Plains. *Agronomy Journal* 97: 364–372.

Table 3.6. Hudson, Berman (1994). *Journal of Soil and Water Conservation* 49 (2) 189–194.

Table 3.7. Nielsen (1995). Water use/yield relationships for Central Great Plains crops. *Conservation Tillage Fact Sheet #2-95.* Akron, Colorado, Central Great Plains Research Station.

Table 3.8. Quarles (1994). Effects of ten years of continuous conservation tillage crop production and infiltration for Missouri claypan soils. *UMC Agronomy Technical Report* 12: 12.

Table 3.9. Moebius (2008). Long-term effects of harvesting maize stover and tillage on soil quality. *Soil Science Society of America Journal* 72: 960–969.

Tables 3.10 and 3.11. Barber (1979). Corn residue management and soil organic matter. *Agronomy Journal* 71: 625–627.

Table 3.12. Top: Blanco-Canqui, et al. (2006). Corn stover impacts on near-surface soil properties of no-till corn in Ohio. *Soil Science Society of America Journal* 70: 266–278.

Bottom: Blanco, Mikha, Presley, Claassen (2011). Addition of cover crops enhances no-till potential for improving soil physical properties. *Soil Science Society of America Journal* 75(4): 1471–1482.

Table 3.13. Ketcheson & Beauchamp (1978). Effects of corn stover, manure, and nitrogen on soil properties and crop yield. *Agronomy Journal* 70: 792–797.

Table 3.14. Magdoff & Amadon (1979). Yield trends and soil chemical changes resulting from N and manure application to continuous corn. *Agronomy Journal* 72: 161–164.

Table 3.15. Moebius (2008). Plant available water capacity vs. cropping system in New York. *Agronomy Journal* 72: 960–969.

Table 3.16. Nichols & Toro (2010). A whole soil stability index (WSSI) for evaluating soil aggregation. *Soil & Tillage Research* 111 (2011): 99–104.

Table 3.17. Basso, et al. (2012). Assessing potential of biochar for increasing water-holding capacity of sandy soils. *GCB Bioenergy* 5: 132–143.

Table 4.1. Chen & Weil (2010). Penetration of cover crop roots through compacted soils. *Plant and Soil* 331: 31–43.

Table 4.2. Burton, Prine, Jackson (1957). Studies of drouth tolerance and water use of several southern grasses in Tifton, GA. *Agronomy Journal* 49: 498–503.

Table 6.1. Begg & Turner (1976). Crop water deficits. *Advances in Agronomy* 28:161–217.

Table 6.2. Crider (1955). Root growth stoppage resulting from defoliation of grass. *USDA Bulletin* 1102.

Tables 6.3, 6.4, and 6.5. Launchbaugh (1957). Bulletin 394, *The Effect of Stocking Rate on Cattle Gains and on Native Shortgrass Vegetation in West-Central Kansas.* Kansas State University Experiment Station.

Table 6.6. Stoddart (1946). UAES Bulletin No. 324, *Some Physical and Chemical Responses of* Agropyron spicatum *to Herbage Removal at Various Seasons.* Utah Agricultural Experiment Station.

Tables 6.7 and 6.8. Owensby, Paulsen, McKendrick (1970). Effect of burning and clipping on big bluestem reserve carbohydrates. *Journal of Range Management.* September 1970, 358–362.

Table 7.1. Ward, et al. (1982). Ammonia treatment of wheat straw. *Beef Cattle Report,* University of Nebraska MP43:12.

Tables 7.2–7.6. Data taken from class materials in Range Livestock Management taught by Dr. Robert Cochran, Kansas State University, 1989.

Table 8.1. David Kraft (2013) USDA-NRCS Kansas Rangeland Management Specialist.

Table 8.2. Leng, et al. (2012). Effect of biochar on rumen methane emissions. *Livestock for Rural Development* 24 (6).

Table 8.3. Leng, et al. (2012). Biochar reduces enteric methane and improves growth and feed conversion in local "Yellow" cattle fed cassava root chips and fresh cassava foliage. *Livestock for Rural Development* 24 (11).

Table 9.1. Schroyer, et al. (1997). *Wheat Production Handbook,* Publication C-529. Kansas State University Extension.

ARTICLES CITED

Page 50: Klein, Bob; Cody Creech; and Rodrigo Werle. "The Value of Wheat in a Crop Rotation." University of Nebraska *CropWatch* bulletin, August 31, 2016. cropwatch.unl.edu/2016/value-wheat-crop-rotation

Page 64: In an Australian study, biochar was mixed with molasses and fed directly to cows. Stephen Joseph, et al. (2015). "Feeding Biochar to Cows: An Innovative Solution for Improving Soil Fertility and Farm Productivity." *Pedosphere*, 25(5): 666–679.
www.sciencedirect.com/science/article/pii
/S1002016015300473

Page 69: Here are several studies of subsoiling.

In University of Kentucky Extension work done by Lloyd Murdock, corn after subsoiling yielded 161.6 bushels per acre, while corn without subsoiling yielded 160.3 — hardly an economical response to such an expensive operation. Murdock, Lloyd W. (1999). "Subsoiling of No-Tilled Corn." University of Kentucky UKnowledge.
uknowledge.uky.edu/cgi/viewcontent.cgi?article
=1004&context=pss_views

In an extensive study conducted in Iowa and Illinois, subsoiling failed to improve yields over multiple sites and multiple years. *See* W. E. Larson, W. G. Lovely, J. T. Pesek, and R. E. Burwell (1960). "Effect of Subsoiling and Deep Fertilizer Placement on Yields of Corn in Iowa and Illinois." *Agronomy Journal* 52: 185-189.
dl.sciencesocieties.org/publications/aj/abstracts
/52/4/AJ0520040185

Research by Voorhees at the USDA-ARS in Waseca, Minnesota, showed not only no improvement due to subsoiling but also an 11-bushel decrease in corn yield over the year the trial was conducted. Jodi DeJong-Hughes, "Will Subsoiling Increase Crop Yields in Minnesota?" University of Minnesota Extension Service, January 2003.
www.extension.umn.edu/agriculture/crops-research/
north/2002/docs/2002-sub-soiling-in-minnesota.pdf

Page 74: Research was conducted at the University of Wisconsin–River Falls. Hankes and Anderson, *The Effects of Mycorrhizal Inoculation on the Drought Stress Tolerance of Corn*, 2006.

Page 125: A summary of 24 trials appeared in *Journal of Animal Science,* March 1986, 62(3): 583-592.

Page 128: Regarding compaction, there have been literally hundreds of studies done by university researchers that indicate there is minimal, if any, yield loss from grazing cropland during the winter. One of the best: Wilson, Casey B.; Erickson, Galen E.; Klopfenstein, Terry J.; Rasby, Richard J.; Adams, Don C.; and Rush, Ivan G. (2004). "A Review of Corn Stalk Grazing on Animal Performance and Crop Yield," *Nebraska Beef Cattle Report,* 215.

Page 134. Guidelines for making a salt-limited supplement: Berger & Rasby, Limiting Feed Intake of Salt in Beef Cattle Diets. *NebGuide*, G2046. University of Nebraska–Lincoln Extension, Institute of Agricultural and Natural Resources.
extensionpublications.unl.edu/assets/pdf/g2046.pdf

Page 163: Crop rotation decisions depend on soil moisture conditions, as demonstrated in this ARS/USDA study. Francisco Calderon, "Adaptation of Dryland Cropping Systems for the Central Great Plains Region to Extreme Variation of Weather and Climate."
www.ars.usda.gov/plains-area/akron-co/cgprs/

Page 165: The Poppers' original essay on the Buffalo Commons can be found online at www.lacusveris.com.

RESOURCES

INDIVIDUALS & ORGANIZATIONS

COLLINS GRAZING

Keyline design and grazing management
802-782-1883
abenewsoil@gmail.com
www.vtfarmtoplate.com/organization/collins-grazing

GRASSWORKS MANUFACTURING

Weed wipers and other products
grassworksmanufacturing.com
888-80-WIPER

NO-TILL ON THE PLAINS

Education and networking on agricultural
production systems that model nature
785-210-4549
swaffar@notill.org
www.notill.org

QUORUM LABORATORIES

Haney Soil Test
844-273-2005
contact@quoruml.com
www.quoruml.com

REGRARIANS

Keyline design and permaculture
regrarians@gmail.com
www.regrarians.org

WARD LABORATORIES

Haney Soil Test
800-887-7645
SampleData@wardlab.com
www.wardlab.com

PRODUCTS

AMAFERM

www.amaferm.com

BATT-LATCH

MSF Farm
http://msffarm.com/fence-products/batt-latch.html

BOVA-PRO

AgroChem
www.agrocheminc.com/bovapro-iodine-concentrates

GREEN COVER SEED

www.greencoverseed.com

K-LINE IRRIGATION

www.k-linena.com

LACTIPRO

MSBiotec
http://msbiotec.com

MYCOAPPLY

Mycorrhizal fungi
www.mycoapplycertified.com

RECOMMENDED READING AND FILMS

BOOKS

Davis, Walt. *How to Not Go Broke Ranching: Things I Learned the Hard Way in Fifty Years of Ranching*. CreateSpace Independent Publishing Platform, 2011.

Gerrish, Jim. *Kick the Hay Habit*. Green Park Press, 2010.

Gerrish, Jim. *Management-Intensive Grazing*. Green Park Press, 2004.

Voth, Kathy. *Cows Eat Weeds: How to Turn Your Cows into Weed Managers*. 2011. (*See also* www.Behave.net)

Yeomans, P. A. *Water for Every Farm: Yeomans Keyline Plan*. CreateSpace Independent Publishing Platform, 2008.

DOCUMENTARIES

The Dust.Bowl
A film by Ken Burns
www.pbs.org/kenburns/dustbowl

Water Movement in Soils
A film by Walter Gardner at Washington State University. Original 1959 and updated 2014 versions can be found by searching on YouTube or in various state university libraries.

METRIC CONVERSION CHART

WHEN THE MEASUREMENT GIVEN IS	TO CONVERT IT TO	MULTIPLY IT BY
inches	centimeters	2.54
feet	meters	0.305
mils	millimeters	0.254
square feet	square meteres	0.093
ounces	grams	31.1
pounds	kilograms	0.373
tons	metric tons	0.907
gallons	liters	3.785

To convert Fahrenheit to Celsius, subtract 32 from Fahrenheit, multiply by 5, then divide by 9.

GLOSSARY

2,4-D. An herbicide that acts on plant hormonal systems, usually used to selectively kill broadleaf plants without harming grasses.

open cows. Cows that are not pregnant.

acre-inches. An amount of water equivalent to a one-inch depth over an acre of land. Since an acre is 43,560 square feet, and an inch is 1/12 of a foot, an acre-inch is 3,630 cubic feet, or 102,790 liters

adsorb. strongly attach

aerenchyma. A porous tissue found in the roots of certain plants that conducts oxygen to the root system, enabling the roots to survive in low-oxygen environments

aggregate. *See* soil aggregate

ammoniation. A process in which low-quality forage is sealed in an airtight layer of plastic and infused with ammonia gas for a period of time. This process breaks the chemical bonds in plant fibers, making the forage more digestible.

auxin. A plant hormone that controls many plant growth responses

bale grazing. A system of feeding hay in which large round bales of hay are unrolled and rationed out gradually with portable electric fence to ensure even distribution of manure and wasted hay across the landscape

Batt-Latch. A device that automatically opens a spring-loaded electric gate at a predetermined time

biochar. Charcoal that has been infused with moisture, plant nutrients, and microbes, used as a soil amendment to improve plant growth

broadbase terrace. A ridge-shaped landform constructed to slow downhill flow of water on agriculture lands, built wide enough to easily be driven over with normal agricultural machinery

chem-fallow. A farming system in which land is laid idle for a period of time to store moisture for a subsequent crop, and plant growth is prevented by herbicides, rather than by tillage

contour tillage. Tillage that creates furrows perpendicular to the direction of the slope, which slows runoff of water, rather than up and down the slope

cyanobacteria. A class of aquatic microbes, also called blue-green algae, capable of producing byproducts that are toxic to animals and humans drinking the water that contains them

fallow. A period of time in which all plant growth is prevented on a field, to store moisture for a subsequent crop

fish trap gate. A gate shaped like a funnel, through which livestock can easily pass in one direction but are prevented from passing in the opposite direction because the gate catches on their shoulders

forbs. Broadleaf flowering (non-grass) plants in rangelands

glomalin. A substance secreted by mycorrhizal fungi that is a powerful soil-improving agent

growing points. Regions on a plant from which new growth originates

hyphae. Root-like structures found on fungi

inversion tillage. Tillage methods that involve the topsoil being placed below the soil surface, and lower soil layers being brought to the soil surface; examples include moldboard plowing and disc plowing

keyline. An imaginary line connecting all points on the landscape at the same elevation as the **keypoint**

keyline design. A system of land-forming designed by P. A. Yeomans to retain rainwater on the landscape, and to move rainwater from wet areas toward dry areas

keypoint. A location within a valley floor where the slope changes from steep to gentle; usually the highest point in the valley where water can be economically impounded

level-bench terrace. A terrace system that carves level areas out of a slope to alternate with vertical risers, giving the landscape a "stairstep" appearance

lodge/lodging. The breaking and falling over of crop stems, which makes harvest difficult

macropore. A soil pore large enough for air and water to move freely through it

management-intensive grazing. A rotational grazing system based on frequent moving of the livestock to new paddocks, with the timing of those moves based upon reaching target levels of defoliation, and rest periods based upon reaching a targeted level of recovery

methemoglobin. Hemoglobin (an oxygen-carrying compound in animal blood) that has been chemically altered (usually by intake of dietary nitrates) in a way that prevents it from carrying oxygen

methanotrophs. Microbes that derive energy from consuming methane

mycorrhizal fungi. Beneficial fungus that live symbiotically upon crop roots and enhance the ability of roots to acquire water and nutrients

no-till. A system of crop farming which eliminates tillage of the soil

ollas. Unglazed ceramic jars that are buried in soil and filled with water, slowly releasing moisture into the surrounding soil; a simple but highly efficient irrigation method

overyielding. A phenomenon that occurs when a mixture of two or more plant species produces more yield of seed or forage than the average of the yield of the component species in pure stands

pasture cropping. A system in which annual grain crops are seeded into dormant perennial pastures to provide both a grain crop and a grazing crop from the same land area within a year

platy layer. A dense soil layer in which normally round soil particles have been compressed into flat, overlapping structures resembling dinner plates. Generally caused by repeated tillage, usually moldboard plowing

plowpan. A dense, hard layer of soil, usually found at a depth of four to six inches deep, created by repeated tillage, particularly by moldboard plowing

photorespiration. A phenomenon that occurs in some plants, particularly cool-season grasses, in which the enzyme responsible for obtaining carbon dioxide for photosynthesis mistakenly acquires oxygen instead, and causes a reduction in photosynthesis and plant growth. Occurs during warm and sunny conditions

pore. A void between particles in the soil

pugging. Churning of the soil by livestock hooves in wet conditions so that pasture plants are damaged

retention dam. A small structure in a stream designed to slow stream flow, create small impoundments, and increase the infiltration of stream water into the soil profile

saturation. A condition in which all voids in a soil are completely filled with water

slake test. A test in which chunks of soil are placed on a screen and submerged in a column of water to determine the rate at which the soil chunks slake (fall apart). A resistance to slaking indicates that a high level of desirable biological glue is present in the soil, and that the soil has good aggregate stability.

soil aggregate. A soil particle composed of many smaller particles of sand, silt, and clay, held together by biological glues

spaced bale feeding. A system of feeding large round hay bales in which the bales are placed on a grid pattern and rationed out to livestock with moveable electric fence, so that manure and trampled hay are uniformly distributed

stalk rot. A fungal disease of grain crops that infests the stems, which reduces the ability of the plant to take up water and nutrients, and causes the stems to break and the crop to fall over

stocker animals. Animals that are retained after weaning and pastured until they reach a heavier sale weight

stocking rate. The number of animals per unit of land area

stomata. Holes in the leaf surface of plants, through which plants take in carbon dioxide and release oxygen to the atmosphere

stover. Crop residue

swale. An elevated landform placed perpendicular to the direction of the slope, designed to impound runoff of rainfall and hold it until it soaks into the soil

terrace. *See* **broadbase terrace** and **level bench terrace**

threshold water use. The amount of plant water use necessary to reach the first unit of grain yield

transpiration. The loss of water vapor from plant leaves and stems

transpiration ratio. The amount of water vapor lost to the atmosphere compared to the amount of plant growth produced

vertical mulch. A narrow trench in the soil filled with coarse organic material, designed to improve water infiltration

water-stable aggregate. A large, composite particle of sand, silt, and clay, held together with biological glues, which is resistant to rapid decomposition when immersed in water

weed wiper. A device consisting of an absorbent material soaked in herbicide, designed to apply that herbicide only to weeds touched by the material

Glossary

INDEX

Page numbers in *italic* indicate drawings and photographs.
Page numbers in **bold** indicate tables and charts.

GROW SMARTER WITH MORE BOOKS FROM STOREY

BY JOSH VOLK

Design a profitable farming enterprise on 5 acres or less! Detailed plans from 15 real farms across North America show how to maximize productivity while minimizing struggles, regardless of whether your plot is rural or urban.

BY ANN LARKIN HANSEN

This comprehensive guide to starting or transitioning to an organic farm equips you with the information you need to earn organic certification for your produce, grains, meat, and dairy.

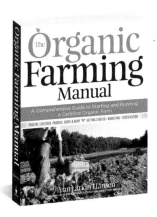

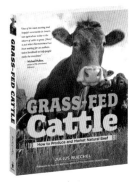

BY JULIUS RUECHEL

Discover the sustainability and great taste of grass-fed beef! This complete manual on raising, caring for, and marketing grass-fed cattle has detailed information on herd selection, pasture management, winter grazing, equipment, slaughter, and much more.

BY DAVID A. BAINBRIDGE

Tackle any drought with these inexpensive, low-tech methods for using water more efficiently in your garden. Illustrated, step-by-step instructions teach you how to deliver water directly to roots, reduce weed growth, and harvest rainwater.

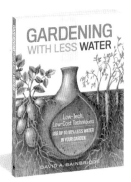

31901063441127